7 DAY

University of Plymouth
Charles Seale Hayne Library
Subject to status this item may be renewed
via your Voyager account

http://voyager.plymouth.ac.uk
Tel: (01752) 232323

LARVAE

MORPHOLOGY, BIOLOGY AND LIFE CYCLE

ANIMAL SCIENCE, ISSUES AND PROFESSIONS

Additional books in this series can be found on Nova's website
under the Series tab.

Additional e-books in this series can be found on Nova's website
under the E-book tab.

MICROBIOLOGY RESEARCH ADVANCES

Additional books in this series can be found on Nova's website
under the Series tab.

Additional e-books in this series can be found on Nova's website
under the E-book tab.

LARVAE

MORPHOLOGY, BIOLOGY AND LIFE CYCLE

KIA POURALI

AND

VAFA NIROOMAND RAAD

EDITORS

Nova Science Publishers, Inc.
New York

For permission to use material from this book please contact us:
Telephone 631-231-7269; Fax 631-231-8175
Web Site: http://www.novapublishers.com

NOTICE TO THE READER

The Publisher has taken reasonable care in the preparation of this book, but makes no expressed or implied warranty of any kind and assumes no responsibility for any errors or omissions. No liability is assumed for incidental or consequential damages in connection with or arising out of information contained in this book. The Publisher shall not be liable for any special, consequential, or exemplary damages resulting, in whole or in part, from the readers' use of, or reliance upon, this material. Any parts of this book based on government reports are so indicated and copyright is claimed for those parts to the extent applicable to compilations of such works.

Independent verification should be sought for any data, advice or recommendations contained in this book. In addition, no responsibility is assumed by the publisher for any injury and/or damage to persons or property arising from any methods, products, instructions, ideas or otherwise contained in this publication.

This publication is designed to provide accurate and authoritative information with regard to the subject matter covered herein. It is sold with the clear understanding that the Publisher is not engaged in rendering legal or any other professional services. If legal or any other expert assistance is required, the services of a competent person should be sought. FROM A DECLARATION OF PARTICIPANTS JOINTLY ADOPTED BY A COMMITTEE OF THE AMERICAN BAR ASSOCIATION AND A COMMITTEE OF PUBLISHERS.

Additional color graphics may be available in the e-book version of this book.

Library of Congress Cataloging-in-Publication Data

ISBN: Larvae : morphology, biology, and life cycle / editors, Kia Pourali and Vafa Niroomand Raad.
     p. cm.
   Includes index.
   ISBN 978-1-61942-662-7 (hardcover)
  1. Larvae. 2. Marine fishes--Larvae. 3. Invertebrates--Larvae. 4. Larvae--Behavior. 5. Larvae--Microbiology. 6. Larvae--Life cycles. I. Pourali, Kia. II. Raad, Vafa Niroomand.
   QL639.25.L35 2011
   571.8'76--dc23
                          2011051314

Published by Nova Science Publishers, Inc. † New York

CONTENTS

PREFACE

Larvae are the distinct juvenile form many animals undergo before metamorphosis into adults. Animals with indirect development such as insects, amphibians, or cnidarians typically have a larval phase of their life cycle. In this book, the authors present current research in the study of the morphology, biology and life cycle of larvae. Topics include the morphological changes of marine fish larvae and their nutrition needs; spawning and nursery areas of neotropical fish species; larvae and embryonization principles of ontogeny in arthropods; larvicidal lectins and trypsin inhibitors isolated from different plant tissues; aquaculture production and larval rearing; expression of recombinant proteins in insect larvae; and adaptive plasticity and evolution in larvae with particular reference to the chironomid midge.

Chapter 1 - Fish larvae represent a critical period in their life history because during ontogenetic development, important changes in structure and function occur in body tissues, organs and systems. Fish larval nutrition starts from an endogenous origin where larvae derive nutrition from yolk reserves and subsequently develop into exogenous feeding. In the early feeding stage, fish experience mixed nutrition during the transitional period from endogenous to exogenous nutrition. In mixed feeding, nutrition supply plays an important role in fish survival. Therefore, an understanding of the structural and functional changes of the larval digestive system is essential to select appropriate feed to improve fish survival in larval rearing. In aquaculture, high fish mortality usually occurs in the early stage that coincides with the period of initial exogenous feeding. Specifically, the timing of food introduction is critical because food ingestion in first-feeding larvae determines subsequent fish survival and growth. In this chapter, we use the life cycle of larval fish as a framework to examine internal factors regulating ontogenetic development in fish larvae and environmental factors contributing to fish mortality during early life history.

Chapter 2 - A hypothetical initial pattern of arthropod ontogenesis, the origin and role of the larval stage and significance of metamorphosis in the life cycle are analyzed and discussed. Arthropods are characterized by the following characteristics: (1) a unique chitinous segmented exoskeleton that is replaced by a new one in every moultingmolting process, and (2) by multi-stage ontogenesis, which both should be treated in tight interrelations. The existence of several developmental stages is thought to be defined by two evolutionary processes going in opposite directions: embryonization and de-embryonization. Within a definite arthropod group, these two processes compete. The result of this competition depends on the starting conditions of the whole ontogeny as well as on the ecological and physiological character and background of realization of the entire ontogenesis, in particular, on the preconditions to the enrichment of the egg in yolk. In its extreme expression, the latter factor may potentially result in non-larval mode of development. Nevertheless, even if the eggs are rich in yolk, only larva with accelerated tissue differentiation and, correspondingly, morphologically different from adult organism, hatches. Consequently, the richness of eggs in yolk provides not the process of "embryonalization" [as identified by Schmalghausen (1982)] but only larval life owing to the remaining embryonic yolk. However, another point of view on the nature of arthropod ontogenesis may be proposed. Initially, deeply embryonized animals with non-larval type of development, i.e., directly capable of reproduction, with a single protostage, in the course of evolution have undergone, first, the superstruction of ontogenesis and, second, de-embryonization, which have both resulted in the appearance of a number of successive postembryonic stages with certain preadaptations and, on the other hand, lead to a wide ecological radiation in different groups. Thus, moultings and changing of stages could have appeared in the ontogenesis of various groups of arthropods not at once but gradually in the course of progressive de-embryonization. The divergence of the developmental ways of the larva and adult animal leads to metamorphosis whichthat, in the most general sense, expresses the transition between the larval and adult developmental programs, particularly if the larval organization possesses not only phyletic features like that of the polychaetous trochophore but also adaptive ones like in maggots.

Chapter 3 - Lectins are hemagglutinating proteins that promote cellular responses due to their interaction with glycosylated molecules. Trypsin inhibitors have been described as endogenous regulators of proteolytic enzymes and as storage proteins. Plant lectins and trypsin inhibitors with insecticidal activity may become an alternative to synthetic insecticides that adversely affect the environment and have promoted the emergence of resistant organisms. These proteins exert deleterious effects on larval survival, weight, feeding ability, and development of insects as well as morphology of larvae, pupae and adults. The digestive processes in larval gut are highly active and their integrity is an essential aspect in insect development. Plant lectins and trypsin inhibitors can modulate the activity of digestive enzymes in larval gut. This chapter reports larvicidal lectins and trypsin inhibitors isolated from different plant tissues (e.g. bark, heartwood, leaves and seeds) with potential applications in the control of numerous insect species, including *Achaea janata*, *Aedes aegypti*, *Anthonomus grandis*, *Callosobruchus maculatus*, *Corcyra cephalonica*, *Ephestia kuehniella*, *Helicoverpa armigera*, *Lacanobia oleracea*, *Ostrinia nubilalis*, *Spodoptera littoralis*, *Spodoptera litura* and *Zabrotes subfasciatus*. The changes in morphology of larvae treated with lectins or trypsin inhibitors, as well as the effect of these proteins on larval α-amylase, β-glucosidase, protease and aminopeptidase activities, are also discussed.

Chapter 4 - The anadromous twaite shad, *Alosa fallax fallax*, is a European Union Berne Convention *Natura 2000* species, requiring protection from range states. In Portugal it is classified as "Vulnerable" by the Instituto de Conservação da Natureza using IUCN criteria but still migrates upriver to spawn in several rivers, *e.g.* River Mira. Evidence suggests that this species exhibits homing behavior. This stock-river relationship is important considering that environmental conditions during the embryo-larval period in the freshwater reaches of rivers play an important role in the future of populations, namely through their impact on growth and mortality rates that affect stage duration and survival. Nonetheless, published information on growth and/or mortality of twaite shad larvae is anecdotal. This study was designed to model growth and mortality rates and study their within season variation and examine to what extent short-term changes in environmental and biological factors affect within-year changes in growth rates and relative survival of larvae of twaite shad in the River Mira.

Plankton samples were collected every other week from 28 March until 7 July 2000, the presumed spawning season, at stations in the upper River Mira. Shad larvae were sorted out of the zooplankton. Microstructure of *sagittae* from *c.* 330 post yolk-sac shad larvae, 7.21 to 20.40 mm standard length, was analysed using an image analysis system and an age-length key was prepared. Growth in length of larvae, $g*$, was estimated based on the backcalculated length-at-age using a model-based resampling (bootstrap) approach. Instantaneous daily mortality rates, Z, of larvae were estimated from the ln-transformed form of an exponential model of decline in abundance of adjusted numbers with respect to age (time). The relationship among nine environmental and biological variables, and instantaneous growth and mortality rates of shad larvae was assessed.

Larvae do not grow at the same rate or experience uniform mortality throughout the spawning season. They were estimated to grow at a faster rate earlier in the season when compared to individuals collected later, after mid-May. Early and later-spawned larvae were subject to comparable and declining mortality rates. Seasonal changes in river (water) temperature and rainfall (a proxy of river discharge) as well as variation in crustacean *nauplii* abundance and feeding incidence, coincident with early larval development, are important variables, regulating the growth and mortality of larval twaite shad.

Chapter 5 - Trichinellosis is a zoonotic disease with worldwide distribution caused by intestine nematode of the genus *Trichinella*. Infectivity of this parasite is determined by host immune and inflammatory response to the infection. The murine cellular and humoral immune response to the infection with low doses of larvae of encapsulating (*Trichinella spiralis, T. britovi*) and non-encapsulating (*T. pseudospiralis*) species was studied. Mice were experimentally infected with 10 muscle larvae of the parasite to simulate natural conditions of the infection in rodents, important reservoirs of trichinellosis.

Both *T. spiralis* and *T. britovi* stimulated the proliferation of splenic T and B lymphocytes during the intestinal phase of infection, but *T. spiralis* activated the proliferative response also at the muscle phase, particularly in B cells. Non-encapsulating *T. pseudospiralis* stimulated the proliferation of T and B cells only on day 10 post infection (p.i.) and later at the muscle phase. The numbers of splenic CD4 and CD8 T cells of *T. spiralis* infected mice were significantly increased till day 10 p.i., i.e. at the intestinal phase, and then at the late muscle phase, on day 60 p.i. *T. britovi* infection increased the CD4 and CD8 T cell numbers only on day 30 p.i. Decreased numbers of CD4 and CD8 T cells after *T. pseudospiralis* infection suggest a suppression of cellular immunity. Both encapsulating *Trichinella* species induced the Th2 response (cytokines IL-5, IL-10) at the intestinal phase and the Th2 dominant response at the advanced muscle phase. IFN-γ production (Th1 type) started to increase with migrating newborn larvae from day 15 p.i. till the end of the experiment. IL-5 production was suppressed during the intestinal phase of *T. pseudospiralis* infection. The immune response to *T. pseudospiralis* was directed more to the Th1 response at the muscle phase, the high IFN-γ production was found on day 10 p.i. and it peaked on days 45 and 60 p.i. The low infective dose of *T. spiralis, T. britovi*, and *T. pseudospiralis* induced a late seroconversion in infected mice. Both *T. spiralis* and *T. britovi* did not evoke an increased specific IgM response, which is typical for the acute infection. Only *T. pseudospiralis* induced a higher specific IgM level in the intestinal phase of the infection, till day 30 p.i. *T. spiralis* caused earlier and more intensive specific antibody response (IgG_1, IgG_{2a}, IgG_{2b}), from day 45 p.i, when antigens from newborn and muscle larvae were accumulated, on the contrary to *T. britovi* (IgG_1, IgG_{2a}, IgG_{2b}) and *T. pseudospiralis* (IgG_{2b}), which induced specific antibody response from day 60 p.i.

Knowledge of the functional consequences of species species-specific parasite infection may be used in immunodiagnosis of trichinellosis as well as immunologist immunologic strategies to control the infection.

Chapter 6 - The production of animals through aquaculture practices has narrowed down the dependence on fisheries-derived products. Aquaculture production yields are increasing but the majority of farmed aquatic animals are still represented by freshwater fish such as carp and tilapia. On the other hand, the production of marine organisms is dominated by several species of fish, crustaceans and molluscs; which have been successfully breed in captivity through their whole life cycles. These advances have been in part supported by applied research that has led to development of well-established rearing practices. The continued growth of the aquaculture industry requires high numbers of good quality postlarvae and juveniles produced in commercial hatcheries. Very often, the larval rearing of marine species represents the main bottleneck of the whole aquaculture production process and this is highlighted by the present situation with the production of marine fish. Although hundreds of species of marine fish are viable for farming, the current production accounts to only 3% of the global production of farmed aquatic animals. This figure is in part explained by the current lack of knowledge on the digestive physiology and nutrient requirements for most marine fish larvae. While mollusc larvae entirely depend on phytoplankton, most economically important species of marine crustaceans and fish are strongly dependent on zooplankton during the larval stage. The intrinsic difficulties in producing large amounts of specific live feeds to support the larval rearing of many marine species represents one of the main problems that marine farmers have to cope with. Therefore, intensive research on the ontogeny and physiology of the larval digestive tract is continuously conducted in order to have a better understanding of the larval digestive processes. New findings on the biology and physiology of marine larvae rapidly impacts technical production aspects such as the shape and size of the rearing vessels, the establishment of optimal larval rearing conditions and the larval feeding protocols used to supply live and inert feeds. New research findings assist nutritionists to formulate diets that can successfully replace live preys needed during the critical larval rearing stages.

Chapter 7 - All metamorphosing insects pass through four consecutive stages before reaching their adult form and these stages are known as the egg, the larva, the semipupa and the pupa. The larval stage comprising individuals who have completely different morphology of the other three, since the body is devoid of external appendages and with internally adipose tissue. From the larval stage begins the growth of individuals who shed their skins several times and depending on the insect considered the substages vary from 3 to 5. In bees in general there are five larval stages called L1 to L5, interspersed with each other for four periods of change ending in larval and pupal changes when successful-L5 larvae enter the pupal stage, where individuals have had adult characteristics. Since ants have about four larval stages (L1, L2, L3, L4), but this number can vary from 3 to 6 depending on the species.

In the larvae of ants there are morphological peculiarities in certain species that set them apart from most reported in the literature so far. An interesting fact was observed in the Argentine ant *Linepithema humile*, an invasive species, since they present a dorsal protuberance in the sixth abdominal segment found exclusively in their larvae. Researchers have tried to explain the social function of this structure. One function described was the nutrition. However the morphological results obtained in this study suggested that the dorsal protuberance has no secretory function, since ducts or pore channels were not found, only small folds in the cuticle, which does not indicate a secretion release function. The secretion released is just for the cuticle compounds and does not synthesize other one which would be important in social behavior. Therefore the unique social function that can be attributed to the dorsal protuberance was mechanical or serving like a facilitator making it easier for the workers to transport the immatures through the different environments in the colony.

Chapter 8 - Larvae in the genus *Thitarodes* (Lepidoptera: Hepialidae) are host to *Ophiocordyceps sinensis*, a unique entomophagous fungal parasite that occurs principally in the alpine meadow environment of the Tibetan Plateau and for centuries a widely known and valuable invigorant in traditional Chinese medicine. *T. pui* is mainly distributed at an altitude of 4100 ~ 4650 m in the alpine meadows and alpine shrub meadows of Mt. Segyi La in Tibet. Larvae feed on plants herbaceous roots and humus fragments. Larval development lasts three to four years (about 1095 ~ 1460 d), including 41 ~ 47 d for the egg, 990 ~ 1350 d for the larva, 35 ~ 41 d for the pupa, and 3 ~ 8 d for the adult. Larval growth involves seven to nine instars. Larval head capsule width is the principal parameter to distinguish each larval instar. The ratio of head capsule width to body length and weight decreased with increasing age. The 7th instar pupates into male adults, the 9th instar linto female adults, and the 8th instar larvae into both males and females. Larvae showed an aggregated distribution where larval densities had their greatest concentration between altitudes of 4100 ~ 4650 m. Pupation occurred in the end of April to early May. Adults emerged during the day at any time between sunrise and sunset from late June to mid July. Courtship occurred between 21:00 to 21:30 h followed by mating last from about one to several hours. The ratio of males to females was 1.5:1. The average oviposition was 768 ± 206 eggs per female. The life table of an experimental *T. pui* population was constructed through laboratory rearing. The total survival rate of the experimental population was 2.6%, and the population trend index was 7.95 which indicated that the population size of the next generation will be 7.95 times greater.

Chapter 9 - This chapter provides information on ontogenetic patterns of neotropical fish species distribution in tributaries (Verde, Pardo, Anhanduí, and Aguapeí rivers) of the Porto Primavera Reservoir, in the heavily dammed Paraná River, Brazil, identifying key spawning and nursery habitats. Samplings were conducted monthly in the main channel of rivers and in marginal lagoons from October through March during three consecutive spawning seasons in 2007-2010. Most species spawn in December especially in Verde River. Main river channels are spawning habitats and marginal lagoons are nursery areas for most fish, mainly for migratory species. The tributaries have high diversity of larvae species: a total of 56 taxa representing 21 families, dominated by Characidae. Sedentary species without parental care are more abundant (45.7%), and many long-distance migratory fish species are present (17.4%). Migrators included *Prochilodus lineatus, Rhaphiodon vulpinus, Hemisorubim platyrhynchos, Pimelodus maculatus, Pseudoplatystoma corruscans, Sorubim lima,* two threatened migratory species: *Salminus brasiliensis* and *Zungaro jahu,* and one endangered migratory species: *Brycon orbignyanus*. Most of these migratory species are vital to commercial and recreational fishing, and their stocks have decreased drastically in the last decades, attributed to habitat alteration, especially impoundments. The fish ladder at Porto Primavera Dam appears to be playing an important role in re-establishing longitudinal connectivity among critical habitats, allowing ascent to migratory fish species, and thus access to upstream reaches and tributaries. Establishment of Permanent Conservation Units in tributaries can help preserve habitats identified as essential spawning and nursery areas, and can be key to the maintenance and conservation of the fish species in the Paraná River basin.

Chapter 10 - In Biotechnology, the expression of recombinant proteins is a constantly growing field and different hosts are used for this purpose. Insects from the order Lepidoptera infected with recombinant baculovirus have appeared as a good choice to express high levels of proteins, especially those with post-translational modifications. Lepidopteran insects are extensively distributed in the world. Species like *Bombyx mori* (silkworm) have been explored in Asian countries to produce a great number of recombinant proteins for academic and industrial purposes. Several recombinant proteins produced in silkworms have already been commercialized. On the other hand, species like *Spodoptera frugiperda, Heliothis virescens, Rachiplusia nu* and *Trichoplusia ni* are widely distributed in the occidental world and Europe. The expression of recombinant proteins based on Lepidoptera has the advantage of its low cost in comparison with insect cell cultures. A wide variety of recombinant proteins, including enzymes, hormones and vaccines, have been efficiently expressed with intact biological activity. The expression of pharmaceutically relevant proteins, including cell/viral surface proteins and membrane proteins using insect larvae or cocoons, has become very attractive. This chapter describes the methods for scaling up the protein production using insect larvae as biofactories.

Chapter 11 - In aquaculture fish larvae rearing still represents a considerable challenge because fish larval nutrition is not yet fully understood. The initial larval stages still rely on live feed to survive and growth and live feed have substantial nutritional imbalances. It is crucial to determine the amino acids and fatty acid requirements of fish larvae in order to formulate suitable inert diets. It is possible to estimate larval amino acids requirements by determining the AA profiles from fish larvae carcass. Although this is just a first step it can give a preliminary indication of possible fish AA requirements. Also, if the AA profiles of fish are determined for several ages during fish larval ontogeny is possible to associate certain changes to specific events of fish development such as metamorphosis. Fatty acids are another extremely important nutrient during the first larval stages. Fatty acids are the most important energetic substrate and are crucial to the development of certain organs such as the eye. A change in the fatty acid profile of fish larvae may indicate a higher requirement of a specific fatty acid or its preferential spare. This study aims to compare the differences between the fatty acid and amino acid profiles of two different marine species and associate these changes to certain events occurring during fish larval ontogeny.

Chapter 12 - The theory of evolution is arguable the most influential theory shaping our thinking about biology. Yet it is almost exclusively a theory of adults. It virtually ignores the immature stages of organisms. Here I attempt to make a contribution toward redressing this imbalance, principally by reference to the aquatic larvae of a ubiquitous insect, the chironomid midge. These larvae occur in wide variety of inland waters, significantly including extreme habitats such as transient pools and hence provide good models to illustrate the adaptive response of larvae to a set of identifiable selective pressures. Importantly, the selective pressures experienced by larval stages are quite different to those experienced by adults and hence lead to divergent evolutionary trajectories.

In: Larvae: Morphology, Biology and Life Cycle
Editors: Kia Pourali and Vafa Niroomand Raad

ISBN: 978-1-61942-662-7
© 2012 Nova Science Publishers, Inc.

Chapter 1

MORPHOLOGICAL CHANGES OF MARINE FISH LARVAE AND THEIR NUTRITION NEED

Zhenhua Ma[1], Jian G. Qin[1] and Zhulan Nie[2]*

[1]School of Biological Sciences, Flinders University, Adelaide, Australia
[2]School of Animal Science, Tarim University, Alaer, P.R. China

ABSTRACT

Fish larvae represent a critical period in their life history because during ontogenetic development, important changes in structure and function occur in body tissues, organs and systems. Fish larval nutrition starts from an endogenous origin where larvae derive nutrition from yolk reserves and subsequently develop into exogenous feeding. In the early feeding stage, fish experience mixed nutrition during the transitional period from endogenous to exogenous nutrition. In mixed feeding, nutrition supply plays an important role in fish survival. Therefore, an understanding of the structural and functional changes of the larval digestive system is essential to select appropriate feed to improve fish survival in larval rearing. In aquaculture, high fish mortality usually occurs in the early stage that coincides with the period of initial exogenous feeding. Specifically, the timing of food introduction is critical because food ingestion in first-feeding larvae determines subsequent fish survival and growth. In this chapter, we use the life cycle of larval fish as a framework to examine internal factors regulating ontogenetic development in fish larvae and environmental factors contributing to fish mortality during early life history.

Keywords: Marine Fish Larvae; First Feeding; Digestive System; Feeding Behaviour; Live Food

1. INTRODUCTION

Most marine fish are oviparous and lay eggs into water where external fertilization occurs. Oviparous fish species produce a large number of eggs, and fertilized eggs are then left to develop without parental care [1]. After hatching, these fish larvae carry a large yolk

---

* Corresponding author: Email: Jian.Qin@Flinders.edu.au.

sac that continues to supply nutrition to the larvae for several days until fish further develop swimming and feeding ability. Once the yolk is depleted, larvae must switch to feeding on zooplankton and failure to ingest external food may lead to death due to starvation.

In marine fish aquaculture, the larval stage is the "bottle neck" limiting fingerling supply for grow-out facilities [2]. Success in the culture of fish larvae depends upon the understanding of fish digestive capacity, nutritional requirement, and live food supply [3]. Because some digestive enzymes can be used as indicators to evaluate the functional development of the digestive system of fish larvae [3, 4], it is essential to understand the digestive physiology of fish larvae during early development [5]. Furthermore, the knowledge of digestive development can help determine the time for weaning and selection for artificial diets [3].

The timing to supply feed with appropriate nutritional composition is a key consideration in marine larval fish culture [6, 7]. Before first feeding, yolk sac reserve is the nutrition source for fish larvae in early life. Upon exogenous feeding, nutrition supplies to larvae determine the success of larval fish rearing. The use of live feed organisms is obligated to the success for most marine fish aquaculture [8]. At present, rotifers (*Brachionus* spp.) and brine shrimp (*Artemia* sp.) are the main species of live feeds that are widely used in fish hatchery. Since 1980s, advanced techniques have been developed to enrich live feeds to meet nutritional demand for a specific fish species [9-11]. However, the high cost associated with live feed production has become a major concern in commercial hatcheries. For instance, Le Ruyet et al. (1993) reported that live food (*Artemia* nauplii) account for 79% of the production cost for sea bass juveniles up to 45 days old. In order to decrease the production cost and increase the productivity efficiency, early weaning onto a formulated diet is essential.

In this chapter, we will use the life cycle of fish larvae as a framework to examine internal factors regulating ontogenetic development in fish larvae and environmental factors affecting fish development. To understand the cause of high fish mortality in early life history, we first review factors related to the ontogenetic development and digestive physiology, and then we focus on issues of first feeding of fish larvae in intensive aquaculture. Finally, we discuss management strategies of using live feeds in marine fish hatcheries.

2. ONTOGENETIC DEVELOPMENT AND DIGESTIVE PHYSIOLOGY

Based on fish ontogeny, the life cycle of a fish is divided into five stages (Figure 1): egg, embryo, larvae, juvenile, and adult. Nutrient requirements of all fish species change throughout their life cycle along with morphological and physiological developments of the digestive track and function. Therefore, a good understanding on the variation of the nutritional requirements at each developmental stage is necessary to develop an optimum feeding protocol for larval fish farming.

In most marine teleosts, nutrition during the embryonic phase is derived from yolk reserve. The embryonic period starts from fertilization and ends at the commencement of exogenous feeding. It is divided into three major phases: cleavage egg, embryo, and free embryo [12].

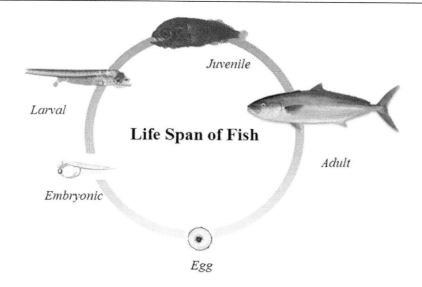

Figure 1. General life cycle of marine fish http://www.dpi.nsw.gov.au/fisheries/recreational /saltwater/sw-species/yellowtail-kingfish, the larval and juvenile images are from our lab, and the embryonic and egg images are adapted from Moran et al. (2007) [26]).

Figure 2 illustrates the embryonic development of yellowtail kingfish (*Seriola lalandi*). Cleavage egg (a-h), embryo (i-l), and free embryo (m) are defined according to Moyle and Cech (2003). Embryonic development is a complex process and egg quality and hatching environments directly affect embryonic development and the sizes of fish at hatching and first feeding [13].

Egg quality is related to broodstock nutrition [14-16] because protein, lipoprotein, glycogen, and enzymes contents in yolk reserve directly affect embryonic development [17-19]. Nutrients such as essential fatty acids (EFA) [20], vitamin E [21], carotenoids [22], vitamin C [23], dietary protein, vitamin B_1, and vitamin B_6 [14] in broodstock diets are essential for the normal development of embryos.

The percentage of normal eggs increases with the increase of *n* -3 highly unsaturated fatty acids (HUFA) in broodstock diets of gilthead seabream (*Sparus aurata*) [20]. Therefore, proper control of broodstock nutrition can improve egg quality and enhance survivorship in marine fish larvae [14].

Ambient environmental factors also affect embryonic development. Higher temperature accelerates development rate [24-28]. For instance, the development rate of mackerel (*Scomber scombrus*) eggs at 17.8 °C is almost three times faster than at 8.6 °C (Figure 3). A similar result has also been reported in striped trumpeter (*Latris lineata*) when compared with incubation temperatures between 16.2 °C and 8.1 °C [24]. However, high temperature over a tolerable range may lead to heat shock and fish mortality [29, 30]. Furthermore, low dissolved oxygen associated with high temperature may contribute embryonic mortality [31, 32]. Severe hypoxia is lethal in the early development of many marine fishes, resulting in fish mortality [32-34]. Generally, the rate of oxygen consumption increases with embryogenesis in fishes such as Atlantic cod *Gadus morhus*, Senegal sole *Solea senegalensis* [35, 36], but it may vary among species.

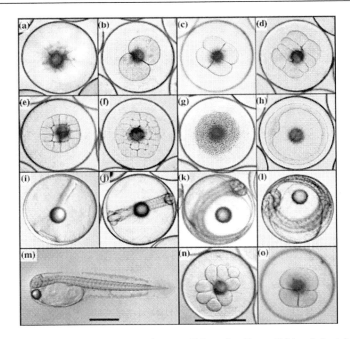

Figure 2. Developmental stages and cleavage abnormalities of yellowtail kingfish: (a) pre-cleavage; (b) 2 cell; (c) 4 cell; (d) 8 cell; (e) 16 cell; (f) 32 cell; (g) mid-stage blastula; (h) gastrula; (i) appearance of embryo; (j) 20 myomere embryo; (k) advanced embryo; (l) pre-hatch embryo; (m) larvae 4 h post-hatch; (n) asymmetrical cleavage in blastula; (o) indistinct cell margins in blastula. Scale bar for (a)-(l) and (n)-(o) shown in (n); scale bars represent 1 mm (Adapted from Moran et al. 2007 [26]).

For instance, the oxygen consumption of clownfish *Amphiprion melanopus* does not increase linearly throughout development [37]. Therefore, temperature and oxygen supply in the incubator can regulate the embryonic development and hatching process in fish larvae.

Newly hatched larvae are a few millimetres long, with different structure, morphology and function from adult [1]. In fish larvae, special larval structure relevant to respiration may develop to increase the area to volume ratio for gas exchange [38]. Fish larvae are delicate and have a large yolk sac and an undeveloped mouth, fins, and eyes (Figure 2m, Figure 4a). The larval stage starts from exogenous feeding and ends with the formation of axial skeleton and the disappearance of median fin-fold [12].

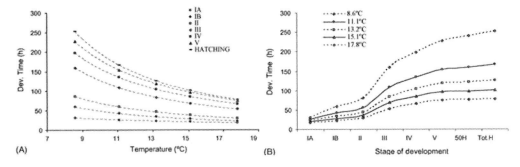

Figure 3. (A) Functional relationship between developmental time, temperature and embryonic stage of mackerel eggs; (B) Fitted relationship between developmental time and temperature at each embryonic stage (Adapted from Mendiola et al. 2006 [182]).

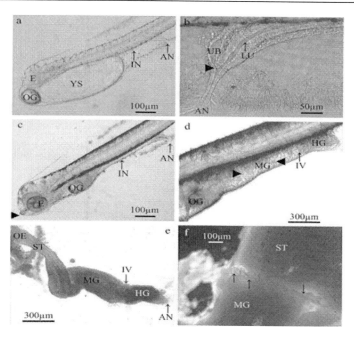

Figure 4. Morphological changes of the digestive tract during the development of yellowtail kingfish larvae. (a) General view of the newly hatched larva on 0 day after hatching (DAH). (b) Posterior region of the incipient intestine of newly hatched larval with a narrow lumen (400×). (c) View of the digestive tract on 2 DAH larvae. (d) Incipient intestine on 4 DAH. (e) Curved gut on 8 DAH. (f) Pyloric caeca (arrows) between the pyloric stomach and anterior midgut on 24 DAH. Abbreviations: AN, anus; E, eye; HG, hindgut; IN, incipient intestine; IV, intestinal valve; LU, lumen; MG, midgut; OE, oesophagus; OG, oil globule; ST, stomach; UB, urine bladder; YS, yolk sac (Adapted from Chen et al. 2006 [57]).

Massive mortality normally occurs during larval stage because of vulnerability of larvae to predation, starvation, unfavourable environmental conditions and prevailing pathogens [12, 39]. In larval fish rearing, fish mortality in the early stage is generally associated with poor food supply under favourable conditions because fish development depends on adequate nutrition uptake [40-43]. After yolk sac is depleted, fish larvae rely on food from exogenous sources [44]. Therefore, the time at first feeding and live food provision is crucial for the growth and survival of postal larvae.

The larval stage constitutes a critical phase throughout the fish life because the ontogenetic development causes important structural and functional changes in body tissues, organs and systems [45]. During this stage, fish larvae undergo a period of mixed nutrition and gradually transfer from endogenous nutrition to exogenous nutrition [6]. Hence, the understanding of ontogenetic development of the digestive system will provide useful knowledge to improve survival and nutrition supplement to fish larvae. The ontogenetic development of the digestive system in teleosts fish is divided into three major phases (Table 1) [57].

In phase I, the primordium of the digestive system appears [46] and the gut forms a straight tube but does not open to the oropharyngeal cavity and anus [47-49] (Figure 4b, c; Figure 5a). This primary digestive system shows little change from mouth opening to the completion of yolk absorption [50]. However, the development pattern of the digestive system is species-dependent. For instance, in some species such as common sole (*Solea solea*), and summer flounder (*Paralichthys dentatus*), the esophagus, incipient stomach and intestine can be distinguished clearly on 3 day post hatching (DPH) [51, 52].

Table 1. Development of digestive system in teleosts fish larvae

Phase	Phase description	Nutrition supply and digestion ability
Phase I	Start from hatching and end at the completion of endogenous feeding;	Yolk sac reserve and oil globule.
Phase II	Start from onset of exogenous and end before the formation of gastric gland in stomach;	Exogenous feeding on live feeds; immature digestive system
Phase III	Starts from the presence of gastric glands and pyloric caeca to metamorphosis.	Artificial pellets; mature digestive system

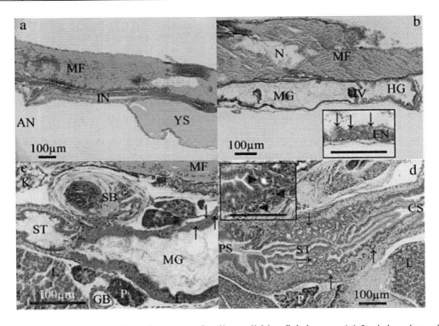

Figure 5. Sagittal section of the digestive tract of yellowtail kingfish larvae. (a) Incipient intestine with narrow lumen on 0 day after hatching (DAH); HE. (b) Larval fish gut on 4 DAH and differentiated hindgut; HE. Inset picture: acidophilic supranuclear vacuoles (arrows) in the hindgut enterocytes (bar = 100μm); HE. (c) Anterior midgut and stomach on 5 DAH. Note the lipid vacuoles (arrows) in the enterocyte; HE. (d) Gastric glands (arrows) in the middle of the stomach on 15 DAH; PAS. Inserted pictures: gastric glands (arrowheads, bar = 100 μm). Abbreviations: CS, cardiac stomach; HE, haematoxylin-eosin; K, kidney; L, liver; MF, muscular fibres; MG, midgut; PAS, periodic acid-Schiff staining technique; PS, pyloric stomach; SB, swim bladder; ST, stomach; Y, yolk residue (Adapted from Chen et al. 2006 [57]).

In this phase, larvae depend on energy reserve in yolk sac and oil globules [53], and these reserves meet the requirement of larval metabolism. Endogenous nutrition at this phase is carried out by endocytosis of the yolk sac and oil globule through a syncytium surrounding the yolk sac [54]. Towards the end of the first phase, fish experience mixed endogenous and exogenous nutrition.

In phase II, fish larvae develop feeding and digestive mechanisms to adapt to exogenous nutrition [55]. As a result of further development of the digestive tract, supranuclear vacuoles and lipid vacuoles appear in the hindgut and midgut (Figure 5 b, c) [56]. The lack of sufficient digestive capabilities is common in this phase [48, 56, 57] and larvae mainly depend on pinocytosis and intracellular digestion and absorption [58]. Pinocytosis is an alternative pathway to digest proteins in teleost larvae as the fish enzymatic system is poorly developed [49]. Digestion takes place in larval intestine without stomach at this stage, where pH remains alkaline and the trypsin-like enzyme activity accounts for the proteolytic activity [59]. At first feeding, both pancreatic and intestinal enzyme activities are generally low [60]. In some species, trypsin, aminopeptidases and alkaline phosphatase are not observed [61], while in other species such as gilthead seabream *Sparus aurata*, Japanese flounder *Paralichthys olivaceus*, and yellowtail kingfish *Seriola lalandi*, phosphatase and trypsin activities are observed 3-4 days after hatching [50, 56]. However, although several digestive enzymes present in the digestive tract, larvae still rely on live food at this stage. As live prey such as rotifers and *Artemia* nauplii not only supply nutrition to fish, but also offer exogenous enzymes for fish larvae to facilitate digestion [62-65]. Besides digestive system, other abiotic factors also affect nutrition acquisition. For example, when fish larvae start exogenous feeding, the size of mouth gape, food particle size, swimming capacity and hunting success also affect food ingestion efficiency [66, 67].

Phase III starts from the presence of gastric glands and pyloric caeca to metamorphosis. The presence of gastric glands (Figure 5d) and pyloric caeca is a sign for digestive system function [52, 68]. After the formation of gastric glands, the detection of pepsin activity is of indicative of stomach function [69]. During this stage, the main enzyme for protein digestion changes from trypsin to pepsin [56], and the digestive system is physiologically ready to process artificial pellets [52, 70]. However, the timing gastric gland formation varies among fish species. For example, in sea bass *Dicentrarchus labrax*, the appearance of gastric gland is on 25 DPH, and in sole *Solea solea* L., the gastric gland occurs on 22 DPH [71]. In contrast, yellowtail kingfish *Seriola lalandi* form gastric glands around 15 DPH, but weaning to formulated pellets can start on 22 DPH [57]. Therefore, the discrepancy of gastric gland formation between species should be considered for the weaning time in different fish species.

3. LARVAL FISH FEEDING

In nature, mortality of fish larvae is 20-99% during the larval period [72] primarily due to predation [73]. Conversely, larval mortalities in intensive aquaculture occur in the early stage primarily due to inappropriate food and feeding [40-43]. Unlike the feeding mode in adult, most fish larvae are predatory planktivores [74].

A typical feeding protocol for larval fish rearing starts from feeding rotifers, and then gradually transfers to *Artemia* nauplii, and finally weans to pellet diets (Figure 6). The period of mixed feeding with rotifers and *Artemia* nauplii is species-dependent.

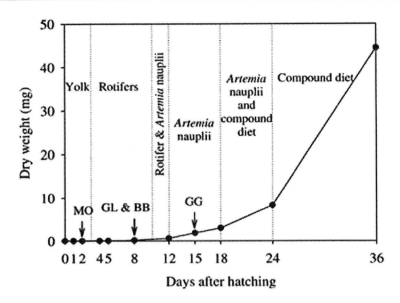

Figure 6. A traditional feeding scheme for yellowtail kingfish larvae as a function of fish weight and day post hatch. Major morphological development events are indicated by arrows: mouth opening (MO) on 2 DAH, formation of gut loop (GL) and brush border (BB) on 8 DAH, and gastric glands (GG) formation on 15 DAH. The diet types at different stages are indicated between the dotted lines (Adapted from Chen et al. 2006 [57]).

3.1. Time of Initial Feeding

The time of initial feeding is critical for survival and growth of fish larvae [75-77]. A short period of food deprivation after yolk absorption can result in poor growth, abnormal behaviour and malformation, alimentary tract degeneration and reduction of food utilisation efficiency [78-80]. Furthermore, when larvae are deprived of food after yolk-sac absorption, fish may permanently lose feeding and digesting ability [39, 40]. Blaxter and Hempel (1963) defined the point of no return for herring as a threshold point during progressive starvation when 50% of the larvae are still alive but too weak to feed, even though food becomes available. This point has also been termed "irreversible starvation" [40]. At the onset of exogenous feeding, fish larvae may die from starvation if the first feeding is delayed and passed the point of no return [40].

It has been demonstrated that starvation coincides with biomass reduction and negative growth [80, 81]. Body energy is consumed at a daily ration of 6-12% body weight, and this evidently produces irreparable damage and mortality when larvae have consumed 35-43% of their caloric content at the point of yolk exhaustion [82]. Therefore, it is vital to detect the exogenous feeding time. Usually, the onset of exogenous feeding in fish larvae can be considered the period from which food ingestion is possible to the point when larval growth is detected [66], and this period usually starts within 24 hours after mouth opening. In practice, swim-up behaviour in fish larvae is a sign to start first time feeding [75, 83, 84].

3.2. Fish Feeding Behaviour

During the period of first feeding, poor swimming ability affects the successful food intake of fish larvae. Hence, it is necessary to understand the swimming and feeding behaviour of fish larvae to maximise feeding success. Fish larvae are classified as either cruising predators or ambush predators [42] (Table 2).

Table 2. Predation mode of selected larval fish species

Cruising fish	Ambush fish
Herring *Clupea harengus* [175]	Atlantic cod *Gadus morhua* [176]
Anchovy *Engraulis mordax* [177]	White crappie *Pomoxis annularis* [176]
Red drum *Sciaenops ocellatus* [178]	Golden shiner *Notemigonus crysoleucus* [85]
Pink clownfish *Amphiprion perideraion* [179]	Striped bass *Morone saxitilis* [180]
Three-spined stickleback *Gasterosteus aculeatus* [181]	Fat snook *Centropomus parallelus* [109]
	Pike *Esox lucius* [181]

A cruising predator moves continuously and searches prey within the boundary of vision space whereas an ambush predator is sedentary and searches prey at the periphery of the strike range [85]. The initial feeding density for an ambush predator should be higher than for a cruising predator. Nevertheless, some fish species may switch search modes depending on prey density, size and distribution [86]. For example, when prey densities decrease, cod larvae (*Gadus morhus*) increase swimming activity and reduce the prey size for selection [87].

3.3. Fish Feeding Ability Associated with Environmental Factors

The ability of fish feeding relies on physical capacities to find, capture, handle and ingest food, and physiological and biochemical capacities to digest and transform the ingested nutrients into energy. However, fish feeding success is also regulated by environmental factors by four mechanisms: metabolic processes, sensory limitations, social interactions and direct impacts [88, 89].

Light can affect the feeding behaviour of fish larvae [88] because most larvae are visual feeders [90]. In addition, light can also affect larval fish feeding by attracting prey to a light source. Light property includes intensity, colour spectrum and photoperiod and these characteristics are extremely variable in nature [91]. Light intensity and spectrum can regulate predator feeding behaviour when searching for prey [92], and photoperiod can stimulate daily feeding rhythms in fish larvae [88]. A long light phase usually stimulates feeding activity whereas a short light phase leads to reduction in feeding for most visual feeding fish.

Temperature is another environmental factor influencing fish development through regulating growth, food intake, food efficiency, and survival [93-95]. High temperature accelerates yolk exhaustion and fish growth but reduces starvation tolerance and survival [96].

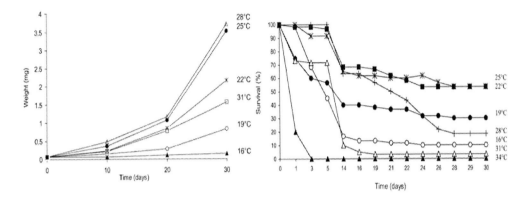

Figure 7. Growth and survival of *Chirostoma estor estor* larvae at different temperatures (Adapted from Martínez-Palacios et al. 2002 [99]).

Food intake increases when temperatures are in the optimal range, falls dramatically beyond the range of optimal temperatures, and fish lose appetite outside the range of temperature tolerance [97]. In general, as temperature increases, morphological development and growth rate increase while survival rate decreases (Figure 7). For species undergoing metamorphosis, the size at metamorphosis increases at low temperatures [98, 99]. High mortality and structural abnormality increases when fish larvae are reared at a temperature outside the optimal range [100-103]. The control of temperature is probably a simple measure to optimise growth and survival when food supply is sufficient [104]. Therefore, for a new fish species, exploring the range of optimal temperatures is a key step for larval rearing success.

Live food density in larval fish rearing affects not only feeding success but also fish production cost. Sub-optimal prey densities during the first feeding usually result in slow growth and high mortality [105]. The empirical concept of larval fish rearing is actually to feed tanks rather than to fish, which usually ends up with high production cost with low feeding efficiency. In a routine hatchery management, larvae are often fed at a much higher prey density than what fish can consume, resulting in food waste and water quality deterioration. The rationale of using a high prey density in larval fish rearing is to improve the encountering rate between prey and predators, and thus increase food consumption [105, 106]. However, the use of extreme high prey density in larval rearing is debatable [107] because there is a disagreement on the notion to constantly expose fish larvae to a high prey density in larval rearing [108-110]. Continuous feeding with a high prey density decreases digestive efficiency and makes the prey passage time in fish gut quicker [111-113]. An ideal larval fish feeding protocol should maintain a high prey density at the initial feeding stage, and then gradually decrease the prey density with increasing visual and swimming ability in fish.

4. LIVE FOOD

According to gut development and yolk absorption, fish larvae are divided into precocial and altricial larvae. Precocial larvae appear as mini-adults, exhibiting fully developed fin and mature digestive system including a functional stomach after yolk sac is fully exhausted. Precocial larvae can ingest and digest formulated diets as a first food (e.g. salmon, trout). On the other hand, altricial larvae remain relatively undeveloped when the yolk sac is exhausted. Most marine fish larvae are altricial, and their digestive system is still rudimentary and lack of a stomach at yolk depletion. Live food is the preferable diet for most marine fish larvae because live feed can stimulate the feeding response of fish larvae and offer exogenous enzymes to help larval digestion [114]. In nature, most marine fish larvae feed on plankton. Therefore, to use a similar foraging environment should be the most reliable way for rearing fish larvae. In intensive rearing, a simple solution is to use live food including unicellular algae to feed filter feeding animals that are in turn fed by fish larvae [13]. Normally, a candidate live food species should be accepted by larval fish and meets their nutrient requirements.

4.1. Algae

Greenwater technology has been widely used in the culture of marine fish larvae [115]. Microalgae can serve as a food source for zooplankton in greenwater tanks and also provide a contrasting background for fish feeding [116]. Therefore, the presence of algae in rearing water usually results in improved growth and survival rate [115, 117]. Although there are numerous hypotheses why greenwater benefits fish larvae in the early stage, the underlying mechanisms are still controversial. These hypotheses include:

(1) Algae in the tank directly or indirectly supply nutrition for fish larvae [118, 119];
(2) Algae may affect the micro-flora in the intestine of fish larvae [116];
(3) Algae in larval tanks may modify the bacterial flora in the water and the live feed [116, 120];
(4) The algae in larval fish tanks may modify the light milieu for larvae [116, 121, 122];
(5) Algae can improve water quality by stripping of nitrogenous substances [122];
(6) Algae may release some compounds acting as attractants to stimulate fish appetite [117]; and
(7) Algae may improve the ability of assimilating micro-particulate diets [118].

In a rearing tank, marine larvae may ingest microalgae, but fish are unlikely to digest the algae. Reitan et al. (1997) suggested that the ingested microalgae might trigger the digestion process or contribute to the establishment of an early gut flora. Cahu et al., (1998) found evidence that algae can trigger digestive enzyme production in pancrease and intestine [123]. Some species such as red drum *Sciaenops ocellatus* may be raised on a micro-particulate diet from first feeding without using zooplankton [118].

With advance in micro-algal production technology, some new products such as instant algae, algae paste, and freeze-dried microalgae are widely available as a substitute for live algae in larval fish rearing. These kinds of algae are grown under laboratory conditions and concentrated for long-life storage. Advantages for using instant algae include stable nutritional profile, more efficient in rotifer enrichment, easy to be consumed by rotifers, and excellent suspension. The use of freeze-dried microalgae in rotifer culture can increase rotifer growth and yield high production [124]. It has been demonstrated no significant difference between using freeze-dried algae and live algae in larval fish growth and survival [125, 126].

4.2. Rotifers

Brachionus rotundiformis (small strain) and *Brachionus plicatilis* (large strain) are the major rotifer species used as live feeds for fish larvae. The wide use of these two rotifers species is because of their fast grow and reproduction, suitable size (80-300 μm) and high density culture rather than being nutritional superior to other live organisms [127]. Rotifers can be fed with different algae, baker's yeast and artificial diets. The densities in rotifer culture range 100-300/ml and the feeding densities to fish larvae are 5-20 rotifers/ml [128]. Typically, rotifers contain 28 - 50% protein and 9 - 28% lipid of the dry weight [115, 127, 129].

Rotifers as the major live feeds for marine fish larvae have been well-known for being low in polyunsaturated fatty acid (PUFA) content [130]. The shortage of PUFA contents in rotifers has resulted in fish slow growth, mass mortality, mal-pigmentation, and deformity [131-135]. In general, the content of EFA such as eicosapentaenoic acid 20:5n-3 (EPA), docosahexaenoic acid 22:6n-3 (DHA) and arachidonic acid 20:4n-6 (ARA) in rotifers are relatively low when compared to other live food (e.g., copepods) in nature [19, 130]. Therefore, enrichment of rotifers with liquid emulsions containing EFA is necessary to improve its quality as fish food [136]. Over the past 20 years, the nutritional aspect of rotifers have received considerable attention [137-141], and various methods for nutritional enrichment of rotifers have been developed [142, 143]. Several commercial products such as Spresso®, DHA Protein Selco®, Algamac 3050®, Ori-gold®, Pav-DHA® are now available on market to enhance not only lipid and protein in rotifers, but also vitamin contents. However, disadvantages also exist by using artificial enrichment emulsions. For instance, oil emulsions may cause rotifer mortalities [144], deteriorate water quality [145], and increase bacterial prolifertion in fish tank [146, 147]. In order to reduce bacterial infection and provide clean rotifers to fish larvae, the use of antibiotics or pro-biotic bacteria is recommended during the period of rotifer enrichment [148, 149]. Despite high production cost, the use of live microalgae such as *Nannochloropsis* sp., *Isochrysis* sp. to enrich live food for fish larvae has also achieved positive results in larval fish rearing [139, 150, 151]. While using live algae for rotifer enrichment, minerals such as iodine and selenium have also been suggested as additives for rotifer enrichment [152].

4.3. Artemia Nauplii

Artemia nauplii have been widely used as a live food following the use of rotifers for larval fish rearing. As there has been no alternative to substitute, *Artemia* nauplii are still an indispensable live food in commercial fish hatchery operations. Nauplii instars I and II are the most widely used forms of *Artemia* in aquaculture. Due to the lack of EPA and DHA, it is necessary to enrich *Artemia* nauplii with oil emulsions before feeding fish larvae. Oil emulsions are general composed of fish oil and fatty acids ethyl esters, providing essential fatty acids to fish larvae in neutral lipid classes [153].

Over the last decades, efforts have been made to improve the formula of enrichment emulsions. In the 1980s, the enrichment strategy was focussed on the use of EPA [154]. In the late of 1980s and early 1990s, the attention was paid to the use of DHA, and the DHA/EPA ratio in the oil emulsion formula [155]. Due to the inherent catabolism, some enriched nutritional components are absorbed by the *Artemia* nauplii after enrichment, resulting in great variation of nutrient contents [156, 157]. Therefore, timely harvesting and storage of enriched *Artemia*, and quick feeding to fish are recommended to minimise nutrition loss after enrichment. Moreover, as some fish species require a high DHA/EPA ratio and more balanced of fatty acid profiles, enriched *Artemia* nauplii may not be a suitable live food. For example, enriched *Artemia* is not appropriate for Pacific bluefin tuna (*Thunnus orientalis*) [158] due to imbalance fatty acid profiles.

4.4. Copepods

Copepods are the major natural food for marine fish larvae in the ocean. Due to superior nutritional value, more and more researchers have focussed on the use of copepods as live food for fish larvae [159-162]. Copepods contain higher essential fatty acids for fish larvae than any other live feeds such as rotifers and *Artemia* nauplii [163, 164]. The nutritional profile of copepods matches the need for fish larval nutrition, especially the content and ratio of PUFA, DHA, EPA, and ARA [165-167]. Positive effects of using copepods to feed marine finfish include improved growth and survival [168-171], body pigmentation [166], and retinal morphology [172]. Although copepod culture in intensive indoor and outdoor systems has been successful, its mass production at a commercial scale has not been achieved due to technical constraints [173, 174].

CONCLUSION

The morphological change is closely related to the nutrition need duing fish early life history. Broodstcok nutrition directly affect egg quality that in turn affects subsequence development of fish larvae. Nutrition supply plays an important role in larval fish survival. Therefore, proper management of broodstock nutrition is primarily important to improve egg quality and fish survorship. During early ontogeny, fish experience endogeneous nutrition, mixed nutrition and exogenous nutrition. Most fish larval mortality in aquaculture occurs in transistional period between endogenous nutiriotn and exogenous feeding. The timing to supply feed to first-feeding larvae is critical to the survival and growth of marine fish larvae. Live food should be supplied to fish larvae within 24 hours after mouth opening. As fish develops, important structural and functional changes occur in body tissues, organs and system. The quality and qanantity of feed supply should change in concomitant with the ontogenetic development of fish larvae. Therefore, the ontogeny of fish digestive track is paramount to understand fish digestive physiology, and this knowledge will provide a guidance for feeding management in larval fish rearing, including live food selection, feeding frequency, and weaning to formulated diet. Environmental factors such as temperature, oxygen and light are fundamental to fish development and nutrition requirement. Particilarly, inappropriate temperatures can retard fish development, induce morphological deformity and mass mortality. Future resarech should be towards (1) the understanding of timing of live food delevery to first-feeding larvae, (2) the nutritional requirments of first-feeding larvae and of fish weaning from live food to formulated diet, (3) the envirorontmal requirments, especially temperature, that are suitable for fish larval surival and growth, and (4) idenficiation of factors causuing fish demormity and mass mortality.

ACKNOWLEDGMENT

The authors tank Liang Song and Xiaoyu Dong for helpful discussion and critical reading of this manuscript.

REFERENCES

[1]   Q. Bone, N. B. Marshall and J. H. S. Blaxter, *Biology of Fishes*, Blackie Academic and Professional, Glasgow (1995).

[2]   J. W. Tucker, *Marine Fish Culture*, Kluwer Academic Publishers, Massachusetts (1998).

[3]   C. A. Alvarez-Gonzalez, M. Cervantes-Trujano, D. Tovar-Ramirez, D. E. Conklin, H. Nolasco, E. Gisbert and R. Piedrahita, *Fish Physiol. Biochem.* 31, 83 (2006).

[4]   X. Shan, Z. Xiao, W. Huang and S. Dou, *Aquaculture* 281, 70 (2008).

[5]   I. Ronnestad, Y. Kamisaka, L. E. C. Conceicao, S. Morais and S. K. Tonheim, *Aquaculture* 268, 82 (2007).

[6]   C. Cahu and J. Zambonino Infante, 200, (2001).

[7]   W. Koven, S. Kolkovski, E. Hadas, K. Gamsiz and A. Tandler, *Aquaculture* 194, 107 (2001).

[8]   B. Baskerville-Bridges and L. Kling, *Aquacult. Nutr.* 6, 171 (2000).

[9]   M. P. Bransden, J. M. Cobcroft, S. C. Battaglene, D. T. Morehead, G. A. Dunstan, P. D. Nichols and S. Kolkovski, *Aquaculture* 248, 275 (2005).

[10]   H. M. Murray, S. P. Lall, R. Rajaselvam, L. A. Boutilier, R. M. Flight, B. Blanchard, S. Colombo, V. Mohindra, M. Yufera and S. E. Douglas, *Mar. Biotechnol.* 12, 214 (2010).

[11]   M. S. Izquierdo, *Cahiers Options Méditerranéennes* 63, 91 (2005).

[12]   P. B. Moyle and J. J. Cech, *Fishes: an Introduction to Ichthyology*, Prentice Hall, Upper Saddle River (2003).

[13]   J. Robin and F. J. Gatesoupe, *Feeding fish larvae with live prey*, in *Nutrition and Feeding of Fish and Crustaceans*, J. Guillaume, S. Kaushik, P. Bergot and R. Metailler, Editors. 2001, Springer, New York. pp213.

[14]   M. S. Izquierdo, H. Fernández-Palacios and A. G. J. Tacon, *Aquaculture* 197, 25 (2001).

[15] C. Mazorra, M. Bruce, J. G. Bell, A. Davie, E. Alorend, N. Jordan, J. Rees, N. Papanikos, M. Porter and N. Bromage, *Aquaculture* 227, 21 (2003).

[16] J. Sawanboonchun, W. J. Roy, D. A. Robertson and J. G. Bell, *Aquaculture* 283, 97 (2008).

[17] R. M. Gunasekera, K. F. Shim and T. J. Lam, *Aquaculture* 134, 169 (1995).

[18] R. M. Harrell and L. C. Woods, *Aquaculture* 133, 225 (1995).

[19] J. Sargent, L. McEvoy, A. Estevez, G. Bell, M. Bell, J. Henderson and D. Tocher, *Aquaculture* 179, 217 (1999).

[20] H. Fernández-Palacios, M. S. Izquierdo, L. Robaina, A. Valencia, M. Salhi and J. Vergara, *Aquaculture* 132, 325 (1995).

[21] H. Fernández-Palacios, M. S. Izquierdo, M. Gonzalez, L. Robaina and A. Valencia, *Aquaculture* 161, 475 (1998).

[22] J. C. A. Craik, *Aquaculture* 47, 61 (1985).

[23] J. H. Blom and K. Dabrowski, *Biol. Reprod.* 52, 1073 (1995).

[24] M. Bermudes and A. J. Ritar, *Aquaculture* 176, 245 (1999).

[25] T. Das, A. K. Pal, S. K. Chakraborty, S. M. Manush, R. S. Dalvi, K. Sarma and S. C. Mukherjee, *Aquaculture* 255, 536 (2006).

[26] D. Moran, C. K. Smith, B. Gara and C. W. Poortenaar, *Aquaculture* 262, 95 (2007).

[27] S. Kazuyuki, K. Hisashi and K. Shogoro, *Comp. Biochem. Physiol. A* 91, 183 (1988).

[28] A. Miranda, R. M. Cal and J. Iglesias, *J. Exp. Mar. Biol. Ecol.* 140, 69 (1990).

[29] S. R. Hopkins and J. K. Dean, *The response of development stages of fundulus to acute thermal shock.*, in *Physiological Ecology of Estuary Organisms*, S. J. Verberg, Editor. 1975, Univ. South Carolina Press, Columbia. pp301-318.

[30] M. Kiyono and K. Shinshima, *Bull. Jpn. Soc. Sci. Fish.* 49, 701 (1983).

[31] F. G. T. Holliday, J. H. S. Blaxter and R. Lasker, *J. Mar. Biol. Ass. U. K.* 44, 711 (1964).

[32] P. J. Levels, R. E. M. B. Gubbels and J. M. Denucé, *Comp. Biochem. Physiol. A* 84, 767 (1986).

[33] Y. Sawada, K. Higuchi, Y. Haga, K. Ura, Y. Ishibashi, M. Kurata, H. Miyatake, S. Katayama and M. Seoka, *Nippon Suisan Gakkaishi* 74, 144 (2008).

[34] K. L. Hassell, P. C. Coutin and D. Nugegoda, *Mar. Pollut. Bull.* 57, 302 (2008).

[35] R. F. Finn, H. J. Fyhn and M. S. Evjen, *Mar. Biol.* 124, 355 (1995).

[36] G. Parra, I. Ronnestad and M. Yufera, *J. Fish Biol.* 55, 205 (1999).

[37] B. S. Green, *Comp. Biochem. Physiol. A* 138, 33 (2004).

[38] E. D. Houde, *Fish Larvae*, in *Encyclopedia of Ocean Sciences*, H. S. John, K. T. Karl and A. T. Steve, Editors. 2001, Academic Press, Oxford. pp381-391.

[39] E. Kamler, *Early life history of fish: An energetics approach*, Chapman and Hall, London (1992).

[40] J. H. S. Blaxter and G. Hempel, *J. Cons. Perm. Int. Explor. Mer.* 28, 211 (1963).

[41] H. Kohno, R. S. Ordonio-Aguilar, A. Ohno and Y. Taki, *Ichthyol. Res.* 44, 267 (1997).

[42] I. Hunt von Herbing and S. M. Gallager, *Mar. Biol.* 136, 591 (2000).

[43] E. D. Houde, *Fish. Bull.* 87, 471 (1989).

[44] X. Shan, H. Quan and S. Dou, *Aquaculture* 277, 14 (2008).

[45] M. I. Sánchez-Amaya, J. B. Ortiz-Delgado, Á. García-López, S. Cárdenas and C. Sarasquete, *Aquaculture* 263, 259 (2007).

[46] J. Zambonino Infante, E. Gisbert, C. Sarasquete, I. Navarro, J. Gutiérrez and C. Cahu, *Ontogeny and Physiology of the Digestive System of Marine Fish Larvae*, in *Feeding and Digestive Function of Fishes*, J. E. P. Cyrino, D. P. Bureau and B. G. Kapoor, Editors. 2008, Science Publishers, Enfield, New Hampshire, USA. pp281.

[47] F. Padrós, M. Villalta, E. Gisbert and A. Estévez, *J. Fish Biol.* 79, 3 (2011).

[48] M. Boulhic and J. Gabaudan, *Aquaculture* 102, 373 (1992).

[49] M. C. Sarasquete, A. Polo and M. Yufera, *Aquaculture* 130, 79 (1995).

[50] S. Kolkovski, *Aquaculture* 200, 181 (2001).

[51] H. Loewe and R. Eckmann, *J. Fish Biol.* 33, 841 (1988).

[52] G. A. Bisbal and D. A. Bengtson, *J. Fish Biol.* 47, 277 (1995).

[53] C. Sarasquete, M. L. Gonzalez de Canales, J. M. Arellano, J. A. Munoz-Cueto, L. Ribeiro and M. T. Dinis, *Histol. Histopathol.* 11, 881 (1996).

[54] T. A. Heming and R. K. Buddington, *Yolk absorption in embryonic and larval fishes*, in *Fish Physiology*, W. S. Hoar and D. J. Randall, Editors. 1988, Academic Press, New York. pp407-446.

[55] H. Segner, R. Roesch, J. Verreth and U. Witt, *J. World Aquacult. Soc.* 24, 121 (1993).

[56] B. N. Chen, J. G. Qin, S. K. Martin, W. G. Hutchinson and S. M. Clarke, *Aquaculture* 260, 264 (2006).

[57] B. N. Chen, J. G. Qin, M. S. Kumar, W. Hutchinson and S. Clarke, *Aquaculture* 256, 489 (2006).

[58] Y. Watanabe, *Bull. Jpn. Soc. Sci. Fish.* 48, 37 (1982).

[59] J. Walford and T. J. Lam, *Aquaculture* 109, 187 (1993).

[60] J. C. B. Cousin, F. Baudin-Laurencin and J. Gabaudan, *J. Fish Biol.* 30, 15 (1987).

[61] H. Segner, R. Roesch, H. Schmidt and K. J. von Poeppinghausen, *J. Fish Biol.* 35, 249 (1989).

[62] S. Kolkovski, A. Tandler, G. Kissil and A. Gertler, *Fish Physiol. Biochem.* 12, 203 (1993).

[63] J. P. Le Ruyet, J. C. Alexandre, L. Thebaud and C. Mugnier, *J. World Aquacult. Soc.* 24, 211 (1993).

[64] A. Gawlicka, B. Parent, M. H. Horn, N. Ross, I. Opstad and O. J. Torrissen, *Aquaculture* 184, 303 (2000).

[65] S. Kolkovski, A. Tandler, G. W. Kissil and A. Gertler, *Fish Physiol. Biochem.* 12, 203 (1993).

[66] M. Yúfera and M. J. Darias, *Aquaculture* 268, 53 (2007).

[67] S. Dou, R. Masuda, M. Tanaka and K. Tsukamoto, *J. Fish Biol.* 60, 1363 (2002).

[68] M. Tanaka, *Jpn. J. Ichthyol.* 18, 164 (1971).

[69] H. Segner, V. Storch, M. Reinecke, W. Kloas and W. Hanke, *Mar. Biol.* 119, 471 (1994).

[70] A. K. Gordon and T. Hecht, *J. Appl. Ichthyol.* 18, 113 (2002).

[71] J. L. Zambonino Infante and C. L. Cahu, *Comp. Biochem. Physiol. C* 130, 477 (2001).

[72] E. D. Houde, *Mortality*, in *Survival strategies in early life stages of marine resources*, Y. Watanabe, Y. Yamashita and Y. Oozeki, Editors. 1996, A.A.Balkema, Rotterdam, The Netherlands. pp51-66.

[73] K. M. Bailey and E. D. Houde, *Adv. Mar. Biol.* 25, 1 (1989).

[74] E. S. Cox and P. M. Pankhurst, *Aquaculture* 183, 285 (2000).

[75] E. Gisbert and P. Williot, *Aquaculture* 156, 63 (1997).

[76] L. Mercier, C. Audet, J. de la Noüe, B. Parent, C. C. Parrish and N. W. Ross, 229, (2004).

[77] M. Yufera, C. Fernandez-Diaz, E. Pascual, M. Sarasquete, F. Moyano, M. Diaz, F. Alarcon, M. Garcia-Gallego and G. Parra, *Aquacult. Nutr.* 6, 143 (2000).

[78] T. A. Heming, J. E. McInerney and D. F. Alderdice, *Can. J. Fish. Aquat. Sci.* 39, 1154 (1982).

[79] B. N. Chen, J. G. Qin, J. F. Carragher, S. M. Clarke, M. S. Kumar and W. G. Hutchinson, 271, 326 (2007).

[80] K. Yoseda, S. Dan, T. Sugaya, K. Yokogi, M. Tanaka and S. Tawada, *Aquaculture* 256, 192 (2006).

[81] J. J. W. Tucker, *Fish. Bull.* 78, 279 (1988).

[82] M. Yufera, A. Polo and E. Pascual, *J. Exp. Mar. Biol. Ecol.* 167, 149 (1993).

[83] D. R. Koss and N. R. Bromage, *Aquaculture* 89, 149 (1990).

[84] T. K. Twongo and H. R. MacCrimmon, *J. Fish. Res. Board Can.* 33, 1914 (1976).

[85] H. I. Browman and J. O'Brien, *Can. J. Fish. Aquat. Sci.* 49, 813 (1992).

[86] R. P. Helfman, *Adaptive variability and mode choice in foraging fishes*, in *Theory and application in fish feeding ecology*, D. J. Stouder, K. L. Fresh and F. R.J., Editors. 1994, University of South Carolina Press, Columbia, South Carolina. pp3-18.

[87] P. Munk, *Mar. Biol.* 122, 205 (1995).

[88] P. Kestemont and E. Baras, *Environmental Factors and Feed Intake: Mechanisms and Interactions*, in *Food Intake in Fish*, D. Houlihan, T. Boujard and M. Jobling, Editors. 2001, Blackwell Science, Cornwall. pp131-156.

[89] A. W. Stoner, *J. Fish Biol.* 65, 1445 (2004).

[90] J. H. S. Blaxter, *Trans. Am. Fish. Soc.* 115, 98 (1986).

[91] G. Boeuf and P.-Y. Le Bail, *Aquaculture* 177, 129 (1999).

[92] A. C. Fermin and G. A. Seronay, *Aquaculture* 157, 227 (1997).

[93] M. Jobling, *Temperature and growth: modulation of growth rate via temperature change.*, in *Gloishbal Warming: Implications for freshwater and marine f*, C. M. Wood and D. G. McDonald, Editors. 1997, Cambridge University Press, Cambridge. pp225-253.

[94] C. A. Martinez-Palacios, M. C. Chavez-Sanchez and L. G. Ross, 2, 455 (1996).

[95] J. R. Brett, *Environmental factors and growth*, in *Fish Physiology*, W. S. Hoar, D. J. Randall and J. R. Brett, Editors. 1979, Academic Press, New York. pp599-675.

[96] S. Z. Dou, R. Masuda, M. Tanaka and K. Tsukamoto, *J. Fish Biol.* 66, 362 (2005).

[97] J. R. Brett and T. D. D. Groves, *Physiological Energetics*, in *Fish Physiology. Bioenergetics and Growth*, W. S. Hoar, D. J. Randall and J. R. Brett, Editors. 1979, Academic Press, New York. pp279-352.

[98] M. Aritaki and T. Seikai, *Aquaculture* 240, 517 (2004).

[99] C. A. Martínez-Palacios, E. Barriga Tovar, J. F. Taylor, G. Ríos Durán and L. G. Ross, *Aquaculture* 209, 369 (2002).

[100] I. Lein, I. Holmefjord and M. Rye, *Aquaculture* 157, 123 (1997).

[101] L. J. Buckley, A. S. Smigielski, T. A. Halavik and G. C. Laurence, *Fish. Bull.* 88, 419 (1990).

[102] G. M. Ludwig and S. E. Lochmann, *N. Am. J. Aqualcult.* 71, 260 (2009).

[103] R. Ørnsrud, L. Gil and R. Waagbø, *J. Fish Dis.* 27, 213 (2004).

[104] C. A. Martinez-Palacios, E. B. Tovar, J. F. Taylor, G. R. Duran and L. G. Ross, 209, (2002).

[105] V. Puvanendran and J. A. Brown, *Aquaculture* 175, 77 (1999).

[106] P. Munk and T. Kiorboe, *Mar. Ecol. Prog. Ser.* 24, 15 (1985).

[107] J. Slembrouck, E. Baras, J. Subagja, L. T. Hung and M. Legendre, *Aquaculture* 294, 52 (2009).

[108] J. Rabe and J. A. Brown, 191, (2000).

[109] S. Temple, V. R. Cerqueira and J. A. Brown, *Aquaculture* 233, 205 (2004).

[110] J. T. Duffy, C. E. Epifanio and J. S. Cope, *J. Exp. Mar. Biol. Ecol.* 202, 191 (1996).

[111] G. W. Boehlert and M. M. Yoklavich, *J. Exp. Mar. Biol. Ecol.* 79, 251 (1984).

[112] T. A. Johnston and J. A. Mathias, *J. Fish Biol.* 49, 375 (1994).

[113] M. F. Canino and K. M. Bailey, *J. Fish Biol.* 46, 389 (1995).

[114] D. A. Bengtson, *Status of Marine Aquaculture in Relation to Live Prey: Past, Present and Future*, in *Live feeds in marine aquaculture*, J. G. Stottrup and L. A. McEvoy, Editors. 2003, Blackwell. pp1-16.

[115] K. I. Reitan, J. R. Rainuzzo, G. Øie and Y. Olsen, *Aquaculture* 118, 257 (1993).

[116] K. I. Reitan, J. R. Rainuzzo, G. Oeie and Y. Olsen, *Aquaculture* 155, 211 (1997).

[117] J. G. Stottrup, K. Gravningen and N. H. Norsker, *ICES Mar. Sci. Symp.* 201, 173 (1995).

[118] J. P. Lazo, M. T. Dinis, G. J. Holt, C. Faulk and C. R. Arnold, *Aquaculture* 188, 339 (2000).

[119] T. Van der Meeren, *J. Fish Biol.* 39, 225 (1991).

[120] J. L. Nicolas, E. Robic and D. Ansquer, *Aquaculture* 83, 237 (1989).

[121] K. E. Naas, T. Naess and T. Harboe, *Aquaculture* 105, 143 (1992).

[122] P. J. Palmer, M. J. Burke, C. J. Palmer and J. B. Burke, *Aquaculture* 272, 1 (2007).

[123] C. L. Cahu, J. L. Zambonino Infante, A. Peres, P. Quazuguel and M. M. Le Gall, *Aquaculture* 161, 479 (1998).

[124] M. Yufera and N. Navarro, *Hydrobiologia* 313/314, 399 (1995).

[125] N. Navarro, M. Yufera and M. Garcia-Gallego, *Hydrobiologia* 452, 69 (2001).

[126] N. Navarro and C. Sarasquete, *Aquaculture* 167, 179 (1998).

[127] A. Srivastava, K. Hamre, J. Stoss, R. Chakrabarti and S. K. Tonheim, *Aquaculture* 254, 534 (2006).

[128] S. Battaglene and S. Fielder, *Hydrobiologia* 358, 1 (1997).

[129] A. V. Frolov, S. L. Pankov, K. N. Geradze, S. A. Pankova and L. V. Spektorova, *Aquaculture* 97, 181 (1991).

[130] L. E. C. Conceição, M. Yúfera, P. Makridis, S. Morais and M. T. Dinis, *Aquac. Res.* 41, 613 (2010).

[131] T. Watanabe, C. Kitajima, T. Arakawa, K. Fukusho and S. Fujita, *Bull. Jpn. Soc. Sci. Fish.* 44, 1109 (1978).

[132] N. Miki, T. Taniguchi, H. Hamakawa, Y. Yamada and N. Sakurai, *Suisanzoshoku* 38, 147 (1990).

[133] T. Takeuchi, J. Dedi, Y. Haga, T. Seikai and T. Watanabe, *Aquaculture* 169, 155 (1998).

[134] M. A. Avella, I. Olivotto, G. Gioacchini, F. Maradonna and O. Carnevali, *Aquaculture* 273, 87 (2007).

[135] I. Olivotto, A. Rollo, R. Sulpizio, M. Avella, L. Tosti and O. Carnevali, *Aquaculture* 255, 480 (2006).

[136] J. R. Rainuzzo, K. I. Reitan and Y. Olsen, *Aquaculture* 155, 103 (1997).

[137] J. Castell, T. Blair, S. Neil, K. Howes, S. Mercer, J. Reid, W. Young-Lai, B. Gullison, P. Dhert and P. Sorgeloos, *Aquacult. Int.* 11, 109 (2003).

[138] C. Estudillo-del Castillo, R. S. Gapasin and E. M. Leaño, *Aquaculture* 293, 57 (2009).

[139] M. Ferreira, P. Coutinho, P. Seixas, J. Fábregas and A. Otero, *Mar. Biotech.* 11, 585 (2009).

[140] W. M. Koven, A. Tandler, G. W. Kissil, D. Sklan, O. Friezlander and M. Harel, *Aquaculture* 91, 131 (1990).

[141] J. Rodriguez Rainuzzo, Y. Olsen and G. Rosenlund, *Aquaculture* 79, 157 (1989).

[142] O. Imada, Y. Kageyama, T. Watanabe, C. Kitajiama, S. Fujita and Y. Yone, *Bull. Jpn. Soc. Sci. Fish.* 45, 955 (1979).

[143] T. Watanabe, T. Tamiya, A. Oka, M. Hirata, C. Kitajima and S. Fujita, *Bull. Jpn. Soc. Sci. Fish.* 49, 471 (1983).

[144] C. Rodriguez, J. A. Perez, M. S. Izquierdo, J. R. Cejas, A. Bolaños and A. Lorenzo, *Aquaculture* 147, 93 (1996).

[145] P. Dhert, G. Rombaut, G. Suantika and P. Sorgeloos, *Aquaculture* 200, 129 (2001).

[146] G. Øie, P. Makridis, K. I. Reitan and Y. Olsen, *Aquaculture* 153, 103 (1997).

[147] A. E. Toranzo, B. Novoa, J. L. Romalde, S. Nunez, S. Devesa, E. Marino, R. Silva, E. Martinez, A. Figueras and J. L. Barja, *Aquaculture* 114, 189 (1993).

[148] R. Verpraet, M. Chair, P. Leger, H. Nelis, O. Sorgeloos and A. De Leenheer, *Aquac. Eng.* 11, 133 (1992).

[149] P. Markridis, A. J. Fiellheim, J. Skjermo and O. Vadstein, *Aquaculture* 185, 207 (2000).

[150] M. Ferreira, A. Maseda, J. Fábregas and A. Otero, *Aquaculture* 279, 126 (2008).

[151] G. M. Ludwig and S. D. Rawles, *J. World Aquacult. Soc.* 39, 158 (2008).

[152] K. Hamre, T. A. Mollan, Ø. Sæle and B. Erstad, *Aquaculture* 284, 190 (2008).

[153] O. Monroig, J. C. Navarro, F. Amat, P. Gonzalez, A. Bermejo and F. Hontoria, *Aquaculture* 251, 491 (2006).

[154] T. Watanabe, C. Kitajima and S. Fujita, *Aquaculture* 34, 115 (1983).

[155] P. Sorgeloos, P. Dhert and P. Candreva, *Aquaculture* 200, 147 (2001).

[156] M. Naz, *Fish Physiol. Biochem.* 34, 391 (2008).

[157] G. V. Triantaphyllidis, P. Coutteau and O. Sorgeloos, *The stability of n - 3 highly unsaturated fatty acids in various Artemia populations following enrichment and subsequent starvation.*, in *Larvi 95 fish and shellfish symposium*, P. Lavens, E. Jaspers and I. Roelants, Editors. 1995, European Aquaculture Society.

[158] A. K. Biswas, J. Nozaki, M. Kurata, K. Takii, H. Kumai and M. Seoka, *Aquac. Res.* 37, 1662 (2006).

[159] G. Dur, S. Souissi, D. Devreker, V. Ginot, F. G. Schmitt and J.-S. Hwang, *Ecol. Model.* 220, 1073 (2009).

[160] T. Camus and C. Zeng, *Aquaculture* 287, 145 (2009).

[161] M. Milione and C. Zeng, *Aquaculture* 273, 656 (2007).

[162] R. Chakrabarti, R. Mansingh Rathore, P. Mittal and S. Kumar, 253, (2006).

[163] J. O. Evjemo, K. I. Reitan and Y. Olsen, *Aquaculture* 227, 191 (2003).

[164] J. G. Støttrup and L. A. McEvoy, *Live feeds in marine aquaculture.*, Blackwell Science Ltd., Oxford (2003).

[165] A. Venizelos and D. D. Benetti, *Aquaculture* 176, 181 (1999).

[166] J. G. Bell, L. A. McEvoy, A. Estevez, R. J. Shields and J. R. Sargent, 227, (2003).

[167] D. A. Nanton and J. D. Castell, *Aquaculture* 163, 251 (1998).

[168] L. A. Copeman, C. C. Parrish, J. A. Brown and M. Harel, 210, (2002).

[169] A. Skalli and J. H. Robin, *Aquaculture* 240, 399 (2004).

[170] M. V. C. Grageda, T. Kotani, Y. Sakakura and A. Hagiwara, *Aquaculture* 281, 100 (2008).

[171] K. Hamre, I. Opstad, M. Espe, J. Solbakken, G.-I. Hemre and K. Pittman, *Aquac. Nutr.* 8, 139 (2002).

[172] R. J. Shields, J. G. Bell, F. S. Luizi, B. Gara, N. R. Bromage and J. R. Sargent, *J. Nutr.* 129, 1186 (1999).

[173] J. G. Stottrup, *Aquac. Res.* 31, 703 (2000).

[174] A. Hagiwara, W. G. Gallardo, M. Assavaaree, T. Kotani and A. B. de Araujo, *Aquaculture* 200, 111 (2001).

[175] H. Rosenthal and G. Hempel, *Experimental studies in feeding and food requirements of herring larvae (Clupea harengus)*, in *Marine food chains*, J. H. Steele, Editor. 1970, University of California Press, Berkeley and Los Angeles. pp344-364.

[176] H. I. Browman and W. J. O'Brien, *Environ. Biol. Fish.* 34, 181 (1992).

[177] J. R. Hunter, *Fish. Bull. US* 70, 821 (1972).

[178] L. A. Fuiman and D. R. Ottey, *Fish. Bull. US* 91, 23 (1993).

[179] D. J. Coughlin, J. R. Strickler and B. Sanderson, *Anim. Behav.* 44, 427 (1992).

[180] J. F. Dower, T. J. Miller and W. C. Leggett, *Adv. Mar. Biol.* 31, 170 (1997).

[181] M. Lehtiniemi, *J. Fish Biol.* 66, 1285 (2005).

[182] D. Mendiola, P. Alvarez, U. Cotano, E. Etxebeste and A. M. de Murguia, *Fish. Res.* 80, 158 (2006).

In: Larvae: Morphology, Biology and Life Cycle
Editors: Kia Pourali and Vafa Niroomand Raad

ISBN: 978-1-61942-662-7
© 2012 Nova Science Publishers, Inc.

Chapter 2

Larvae, Embryonization and Morphophysiological Principles of Ontogeny in Arthropods

Andrew B. Shatrov[*]

Zoological Institute of the Russian Academy of Science,
St. -Petersburg, Russia

Abstract

A hypothetical initial pattern of arthropod ontogenesis, the origin and role of the larval stage and significance of metamorphosis in the life cycle are analyzed and discussed. Arthropods are characterized by the following characteristics: (1) a unique chitinous segmented exoskeleton that is replaced by a new one in every moultingmolting process, and (2) by multi-stage ontogenesis, which both should be treated in tight interrelations. The existence of several developmental stages is thought to be defined by two evolutionary processes going in opposite directions: embryonization and de-embryonization. Within a definite arthropod group, these two processes compete. The result of this competition depends on the starting conditions of the whole ontogeny as well as on the ecological and physiological character and background of realization of the entire ontogenesis, in particular, on the preconditions to the enrichment of the egg in yolk. In its extreme expression, the latter factor may potentially result in non-larval mode of development. Nevertheless, even if the eggs are rich in yolk, only larva with accelerated tissue differentiation and, correspondingly, morphologically different from adult organism, hatches. Consequently, the richness of eggs in yolk provides not the process of "embryonalization" [as identified by Schmalghausen (1982)] but only larval life owing to the remaining embryonic yolk. However, another point of view on the nature of arthropod ontogenesis may be proposed. Initially, deeply embryonized animals with non-larval type of development, i.e., directly capable of reproduction, with a single protostage, in the course of evolution have undergone, first, the superstruction of ontogenesis and, second, de-embryonization, which have both resulted in the appearance of a number of successive postembryonic stages with certain preadaptations and, on the

---

[*] E-mail address: chigger@mail.ru, Fax: +7 812 7140444.

other hand, lead to a wide ecological radiation in different groups. Thus, moultings and changing of stages could have appeared in the ontogenesis of various groups of arthropods not at once but gradually in the course of progressive de-embryonization. The divergence of the developmental ways of the larva and adult animal leads to metamorphosis whichthat, in the most general sense, expresses the transition between the larval and adult developmental programs (Cohen, Massey, 1983), particularly if the larval organization possesses not only phyletic features like that of the polychaetous trochophore but also adaptive ones like in maggots.

Keywords: Metamorphosis, individual development, de-embryonization

INTRODUCTION

In the question on the morphological basis and sources of ontogenesis in arthropods or, otherwise, in the question on the origin and evolution of their ontogenesis as such, there are comeing together many important problems of the modern zoology and, in particular, the problem of the origin and evolution of the arthropods, because, as is widely assumed, the evolution in the broad sense is realized in the form of evolution of ontogeny (phenotypes). The latter is valid not only for arthropods but for all animals (see, for instance, Severtsov, 1970; Kamshilov, 1970, 1972; Schmalghausen, 1982, Ivanova-Kazas, 1997 a, etc.). Concerning arthropods proper, besides a lot of many particular investigations of ontogenesis of representatives of different groups, the main attention has been focused, through the prism of their development and morphology, on their phylogenetic relationships (Sharov, 1965a, б; Anderson, 1973; Manton, 1973; Cisne, 1974; Cutler, 1980; Nielsen, 1985, 1994, 1997, 1998; Weygoldt, 1986; Bergström, 1986, 1992; Starobogatov, 1991; Fryer, 1996, etc.), and on the question of their mono- or polyphily, there are significant differences of opinion, quite formidable on the present level of knowledge. Leaving this controversial question aside of our consideration, let us put, however, our special accent on the key and causal aspects of ontogenesis, which are characteristic not only for arthropods proper but for articulates in general.

In one of the previous works (Shatrov, 1993), the author has viewed ontogenesis mostly from the position of its division into developmental stages and molting processes. It was shown that ontogenesis always evaluates as a whole, and a morphological starting point of each instar is strictly defined by the time of deposition of the very initial layers of the newly formed cuticle during the molting cycle. In the present communication, the main focus directs to the evolutionary significance of the larval stage and also to the problem of metamorphosis.

LARVAE AND MORPHOPHYSIOLOGICAL PRINCIPLES OF ONTOGENY

As is known, the basis of the morphological constitution of arthropods is performed by a unique segmented chitinous cuticle of the exoskeleton of the animal (Nielsen, 1997), and, as a consequence, division of ontogenesis into different developmental stages (instars) strictly marked by molting. These two phenomena are tightly interconnected in historical and individual development, mutually determined, and cannot be considered regardless from one another, because molting processes delimiting existence of instars are mostly caused by the exchange of cuticle and renovation of the organism as a whole. At the same time, in the recent ontogeny, two opposite tendencies of development compete that is are most characteristic for arthropods proper – embryonization and de-embryonization (Zachvatkin, 1975). In this respect, it is most important to realize for how independently, if any, different ontogenetic instars may evaluate (see Severtsov, 1970) and to clarify most possible causes, sources and mechanisms of the ontogenetic processes in their evolution dynamics. The corner stone here is a question on the origin of the larva proper, its primarity and its role in ontogeny and evolution of the different groups that cannot be in no any way underestimated (Cohen, Massey, 1983; Williamson, 1992).

In the examination of the ancestor or sister (Weygoldt, 1986; Fryer, 1996; Nielsen, 1997, 1998) group of arthropods that is thought to be annelids, it appears that larva of polychaete in a pure sense, i.e.,e.i. non-articulated swimming trochophora – it is a real gastrula, embryo, which, however, cannot be attributed in this case as an embryo because this larva lives, moves and develops (Zachvatkin, 1975). In accordance with some ideas (Jezhikov, 1953), in polychaetes, a primary metamorphosis occurs, and trochophora remains in those groups where it is favorable for successful existence of the species. It is important to have in view that in such a character of individual development, nearly everything is taken out from embryogenesis, and further development and complication of the organism are realized already in a living animal during postembryonic development and successive developmental stages (instars). In other words, nearly the whole individual development proceeds not within the indifferent "blastema" of the embryo, but within rather differentiated tissues of the active organism. The latter is especially seen in the case of endolarval mode of development of larvae in the primitive polychaetes (Minichev, 1997). In such an approach, it appears that in the beginnings of ontogenesis of articulates in general and arthropods in particular, there may situate an ultimate form of de-emryonization, as a real phenomenon and not as a process, characterized by a large number of sequential homonomic body segments in a mature organism and, correspondingly, its developmental stages (instars). It is important to note that trochophora may also undergo further de-embryonization (Ivanov, 1937). At the same time, it should be remembered, however, that already in polychaetes, during evolutionary enrichment of eggs in yolk, complication of embryogenesis nevertheless proceeds, i.e., embryonization of the initial stages of development of different intensity occurs up to the situation when a young annelid worm hatches immediately from the egg and development acquires a direct character totally without larva (Zachvatkin, 1975). Thus, even in polychaetes, the whole specter of embryonization, as a phenomenon and so a process, is observed. It is of importance, that trochophora, as individual larval instar, never undergoes any deep independent evolution, and is rather fast quickly (in historical retrospection) embryonized. In arthropods, this developmental stage is totally embryonized, and nauplius of crustaceans is a result of the secondary de-embryonization (Starobogatov, 1991). In general, the primarity even of trochophora – is a rather disputable question.

If addressed to an more earlier and deeper history of life, it appears that, in accordance with the assumption made by K. Nielsen (1985, 1994, 1997, 1998), at the basis of the evolution of the Protostomia and so of Articulata and Arthropoda lies a pelagic simply organized animal of a type of gastrula (recent trochophora), a particular initial larva (protolarva) (the so-called trochea), which, after settling on the ground, transforms slightly and begins to reproduce. There is also an opinion that in the very beginning of the life history, the development was generally and particularly direct without a larva, which, however, evolved only afterwards, and, consequently, without a metamorphosis (Wolpert, 1990). Thus, it appears that the whole primary development was totally focused within only one protostadium, and the whole potential ontogenesis was hidden within the "plastic" mass of the proto-embryo. In this case, there is not any recapitulation in the ontogenesis that is worth to say aboutdiscussing. Concerning similar ideas of Bergström (1986, 1992), at the basis of the whole phylum Bilateria, an endoproctnous larva lies, i.e., a given particularly organized protostadium closed to trochophora. It should be noted, however, that this point of view (the theory of trochea) was subjected to heated criticism (Ivanova-Kazas, Ivanov, 1987) but is still widely spread within the scientific community.

Whatever special organization of the hypothetical initial form was, it should be the following idea should be emphasized the following idea. If recognized that the initial arthropods possessed the least embryonized development, it appears that they obtained at once a complicated ontogenesis in the form of sequential developmental stages (instars) divided by molts and, consequently, a primary metamorphosis. Most likely, however, that sufficiently simply organized and able to reproducetion pelago-bentous animals in the course of the further evolution have undergoingundergone, first, super-building of ontogenesis (lengthening of genome) and, second, de-embryonization, that have lead to the origin of the number of sequential instars, and their change has been marked by molts. Gradual formation of different developmental stages and also their further differentiation have served a causative reason of a wide divergent radiation of different groups of the initial arthropods.

In accordance with a rather widespread point of view (see, for instance, Starobogatov, 1991), any structure or character (for example, body segments or processes) for manifestation in the ontogenesis must yet contain within the embryogenesis;, otherwise, they have never been originated from. But into embryogenesis these characters must firstly somehow be introduced, apparently from (1) still present postembryonic ontogenesis owing to embryonization of the earlier developmental stages (instars) and (2) super-building mutations. From the other hand, that for these characters and, consequently, instars to be in the ontogenesis, they have to be, first of all, super-built over the proto-stadium or have to come from the "plastic" mass of the proto-embryo, i.e., to come from the single initial stage of the proto-arthropod. In other words, development with the single proto-stadium appears to be deeper and more ancient than long ontogenesis, that which is probably also valid for arthropods. In general, instars before going back into embryogenesis have to come from it. Most likely, in each arthropod group there has been initially formed a strict number of body segments and, consequently, instars and molts, and only afterwards during adaptations and specializations, segments and instars previously going out from the proto-stadium, have begun coming back into embryogenesis under protection of developing egg envelops. Later on, it has lead to cephalization and formation of particular pro-morphologies owing to particular adaptations. As a result, a primary larva has disappeared in consequence of embryonization, and changed to a secondary larva of a different advanced level. It appears in such approach that the initial larva of arthropods proper could not ever have been as primary gastrula-like larva but rather more specialized secondary larva of a type of anaprotaspis of trilobites or protonymphon of panthopods (Starobogatov, 1991).

Gradually coming from the proto-stadium or from the primary embryogenesis and radiatinge throughout the ecological niches, the organism of the initial arthropod has required protection in the form of the solid cuticle and also its periodic change with the aim of growth and development, that which have served as a source of stadiality of ontogenesis and have lead to a wide divergent radiation of different arthropod groups. In this case, at the starting point of ontogenesis, there may be only maximum embryonization, where the hatching animal needn't molt many times to achieve mature condition and, consequently, there couldn't be a complex ontogenesis.

In the evolution of ontogenesis of the particular arthropod groups, these two phenomena, embryonization and de-embryonization, probably compete on the basis of already existed ontogenesis, which acquires a fromform of a law (Ivanova-Kazas, 1997), and is due to environmental pressure. In particular, from the point of view of physiological aspects of the realization of particular ontogenesis, a significant role is obviously played here by an initial amount of the yolk in the egg that in due course, is defined by behavioral conditions of mature animals and by the characters of their feeding and oogenesis (Shatrov, 1998). This situation is resultsed in a "'vicious circle'" – a condition of realization of ontogenesis and particular behavioral conditions within the different natural environments, that is directly influencesd on the enrichment of eggs in yolk, in turn influencinge directly on the all characters of the whole further ontogenesis via morphological correlations and inductions. The causative reason of different type of individual development and also the question of realization of particular ontogenesis require in each case special investigations.

During the analysis of the processes of the larva formation, it is important to take into consideration that embryonization and reduction of ontogenesis are first of all a result of the progressive enrichment of eggs in yolk. In this case, a newly hatched organism will look be lesser and lesser looked like the primary larva, but, conversely, would obtain more and more derived character, retaining at the same time the larval status, i.e., the status of non-reproducing instar morphologically different from a mature animal. Reverse process – depletion of eggs in yolk – inevitably may lead to an earlier more early hatching of the organism from the egg, i.e., to de-embryonization. But it is found to be not a very simple problem. As is known from the works of I.I. Schmalghausen (1982), a real "ebryonalization" becomes possible only when the larva as such is totally lost from development, i.e., when it is totally taken up by an embryo. In the presence of any larva, or non-reproducing instar in development, the existence of more or less deep embryonization is impossible. Thus, in the most general case, a larva in the ontogenesis will be expressed in all that times when yolk (or any other) resources of eggs are rather low, i.e., when there are no sufficient internal resources in the embryo to feed developing organism up to the mature reproducing form. At the same time, the intensity and even the presence of embryonization, i.e., the presence of obvious but hidden within the embryonic development early instars is to a great extent debatable that isas equally correct as for arthropods so as for any of groups.

As concerns particular mechanisms of the formation of larva as such in the ontogenesis of arthropods proper, as far as is known from particular investigations and also from the works of O.M. Ivanova-Kazas (1979, 1997), arthropods cannot be characterized as a group (1) of poorly provided of eggs in yolk as well as (2) having extremely small eggs for the exceptions of some specialized, in particular parasitic groups, like, for instance, ichneumonoideas. Nevertheless, quite often, a larva proper is hatched, morphologically different from the adult organism that may be characterized as phenomenon of real de-embryonization carried out in the earliest historical periods. In this situation, the enrichment of eggs in yolk serve for the most part not the processes of "embryonalization" in the sense of Schmalghausen (1982), but only the existence of the larva just after hatching owing to the presence of the embryonic yolk in the mid-gut. In such a case, early differentiation of the tissues and organs may realize in the hatching larvae as is found in some groups of Acari – in a sense of uneven acceleration of some structures on the ground of the extreme small sizes of the organisms proper (McNamara, 1986). This fact may be illustrated on the example of trombiculid mite larvae (Trombiculidae), where mesodermal derivatives – free blood cells haemocytes – differentiate already in prelarva, significantly earlier than the mid-gut undergoes epithelization (Shatrov, 2000).

Larva as a particular stage of hatching, undergoes double effect, – as from the egg and somatic condition of embryogenesis, so from the whole further ontogenesis due to particular morphogenetic correlations by the principle of feedback (Schmalghausen, 1982). At the same time, a large independent role in the evolution of various groups including arthropods is given to larva (Cohen, Massey, 1983; Williamson, 1992) andthat may be also attributed to the whole Metazoa in general (Nielsen, 1997). It should be noted, in particular, that in accordance with the latest assumption (Krivolutskiy, 1999), the Acari – is the pedomorphic, most probably progenetic group, originated in an aquatic environment from larvae of the ancient trilobites even in the Precambrian Era. Surely, this example is not single instance.

In general, in order to say objectively on the question, what ontogenetic processes (embryonization, de-embryonization, various heterochronia, etc.) take place during the developmental processes of various animal groups, including arthropods, it needs is necessary first of all to realize which the character of ontogenesis was the first one for the given arthropod group and what was the starting point from which observable ontogenetic and, consequently, phylogenetic processes have originated (McNamara, 1986). It should be noted, however, that such approach to the historical substantiation of ontogenesis is not shared by some authors (see, for instance, Shishkin, 1988). It is pointed out in this respect theo "…impossibility of direct perusal of ontogenesis as evolution sequence" (Shishkin, 1988, p. 189). At the same time, being remaininged within rather narrow taxonomic groups and relying on the obvious discontinuity of individual stages of arthropod development, it is possible to attempt to understanding causative reasons of evolution origin of their recent ontogenesis. So, if we could ascertain, for instance, that in the initial ontogenesis of all Acari, in contrast with any other groups, there were six developmental stages, or any other but certain number of stages, whatever they transform afterwards, it could inevitably recognized that Acari is a monophyletic taxon as V.B. Dubinin (1959) has supposed, and not polyphyletic as mostly is considered at present. At the same time, it should be noted that Acari in general shows a great conservatism as in morphology so as in the character of development.

Nevertheless, particular examples of individual development of Acari obscure to a great extent general course and tendencies of recent ontogenesis in different groups of these arthropods. B.A. Vainstein (1978), for instance, has reported that some water mites possess unlimited teloblastic growth in their postembryonic development. This phenomenon may indicate a continued coming out of body segments from embryogenesis and, consequently, initial but previously hidden developmental stages. Conversely, in trombiculid mites, a particular very thin and hardly distinguishable cuticle of the hypothetical pre-prelarval instar retains, which is deposited by the embryonic ectoderm cells still before formation of prelarva cuticle (Shatrov, 1999). Thus, an initial number of hypothetical previously active instars in trombiculid mites were not lesser that seven. It is obviously seen from this situation that an apparent immersing into embryogenesis previously existed ining early developmental stages, i.e., a continued process of embryonization.

Concerning prelarva and its evolution and ontogenetic status there is existsed, however, a controversial opinion (Otto, 1997), which considers an active prelarva of a number of acarine groups as a pure embryonic stage obtaining a tendency to a further development. Nevertheless, in such an opinion this author is not the first one and only repeats ideas of many scientists on the nature of metamorphosis (see for reference, Wigglesworth, 1954), which regard all developmental stages nearly up to reproducing one as embryonic. Indeed, if as an embryo, consider all the organisms, which that do exist owing to reserves of nutritive materials in an egg or for a long time livinge within and under protection of the egg shell that are frequently conceptually intermingled but really are not the same, then it appears that not only quiescent or sometimes active prelarva of Acari (Otto, 1997) or analogous foetal instar of many other arachnid groups (Canard, Stockman, 1993) but even active larva retaining embryonic yolk within the mid-gut would be embryonic stages receiving impulse to active life. Recognizing these ideas as rather preposterous, it is necessary to assume, however, that they directly point to the possibility of coming out from embryogenesis a number of stages, i.e., permitting the presence of de-embryonization (Shatrov, 1998).

Applying to ontogenetic evaluation of the arthropod developmental stages, it is necessary to emphasize that *a stage – is a universal, the least from functionally significant, discrete morph of ontogenesis, enclosed within the interval between two successive molts in the combination of the unique morphological characters and adapted to particular environmental conditions.* Although stages, as considered, possess particular evolutionary self-dependence (Garstang, 1922; Severtsov, 1970; Shishkin, 1988), they, however, cannot be able to evaluate quite independently from the remaining ontogenesis, which is always selected and evaluated as a whole (Schmalghausen, 1982). That is way, ontogenesis is represented by "a chain" (series) of interconnected morphological forms, or stages (instars), with their own characters, ontogenic and evolution dynamics of which owing to morphogenetic correlations tightly interconnected with characters and structures of other developmental stages of a given organism (Queiroz de, 1985; Kluge, 1988). Proceeding from this postulate, it is necessary to acknowledge that no one developmental stage can possess the same ontogenetic status. Because cuticle, as is mentioned above, is one of the most fundamental characters of arthropods, and change of cuticle during molting process is the real border between stages, the appearance of the united cuticulin layer – the most superficial and initial layer – of the newly deposited cuticle during molting would manifest the starting point of existence of each new stage in the ontogenesis (Shatrov, 1993).

In the question about ontogenetic status of developmental stages, it is important what we would consider as starting point. It would be doubtfully correct to overemphasize the meaning of the given stages in their strong correlation with the presence or absence of particular structures and organs (Zachvatkin, 1953). In reality, only moltings are valid and important in the context of the ontogenic strategy, as a complex answer on the challenge from particular environmental conditions. If we would manage to prove that in a given ontogenesis there are so many molts, in this ontogenesis are so many stages counted from the first, which could be not homological to the same stage of another ontogenesis. It follows from this assumption that building of phylogenetic trees based on the comparison of only one organ or structure of only one developmental stage rather poorly assist in comprehension of the place of the given group within the given phylum.

METAMORPHOSIS IN ARTHROPODS AND ITS CRITERIA

Metamorphosis is considered ans essential attribute of development of many if not all arthropods, including trombidiform mites (suborder Actinedida) (see, for instance, Vainstein, 1978), so a question of metamorphosis inevitably follows the logic of investigation of their ontogenesis. The problem of metamorphosis is quite complicated because it deals with the conception of development, in which all disciplines of biology from ecological and population to molecular and genetic are focused. There have been issued at least two large collection books on this item – in the years 1968 and 1981 years, in which all of these aspects are just considered, mostly, however, on the example of holometabolous insects and amphibians.

Having an aim to analyze metamorphosis as such, it would be, of course, better first to agree, in which groups metamorphosis is obviously expressed and to observe these groups. But as far as under metamorphosis, there comprehend a rather wide circle of ontogenic processes are comprehended;, it would be expedient first of all to formulate some general approaches and principles of essential parts of metamorphosis and go from general to particular and never by the other way.

Criteria and treatments of metamorphosis are rather different. If proceeding from the most general consideration, any development is in a common sense metamorphic because an organism during its development inevitably changes its appearance, especially if in the ontogenesis there is a larva different from adult form. That is way, it is declared that when germ cells change, – it is evolution, and when a body changes, – it is metamorphosis (Wald, 1981). This sufficiently radical assumption removes all possible questions about the presence of metamorphosis in any animal groups, and particular resolving of these questions becomes a matter of taste of a scientist. However, there are existedexist more special definitions of metamorphosis, which may be summarized as follows. Metamorphosis is a simple changing of the body form during development (Wald, 1981), and a process of transformation of more earlyearlier stage into more later one, i.e., polymorphism enclosed within genome (Highnam, 1981), owing to which an organism irreversibly switches its adaptations from one of its function to the other (Wilbur, 1980). Metamorphosis is also interpreted as a complex process of the intensive morphogenesis during the course of one or two molts and realization of developmental potential of an adult animal, when the individual through the length of life is adapted to existence in more than one ecological niche (Locke, 1981, 1985; Sehnal, 1985). That is way, metamorphosis in the general case reflects transformation of the organism between sufficiently different age-specific adaptive types (Shishkin, 1988). It is not difficult to see that in the later case, the conception about habitat appears in the definition of metamorphosis. In accordance with the author's idea (Shatrov, 1991), metamorphosis – is a radical changing of organization of an individual, mostly determined by redistribution of morpho-functional characters of different organs and closely related with changing of habitat on different developmental stages. In the most general case, as applied to insects in accordance with definition of G.Ja. Bey-Bienko (1980), the essence of metamorphosis is that developmental individual undergoes sufficient reconstruction of its morphological organization and biological peculiarities during its life, and moreover, this author attributed anamorphosis to metamorphosis, i.e., considered metamorphosis in a rather wide sense. There, are of course, existed some other views on the nature of metamorphosis, such as:

metamorphosis is identified with divergent evolution of polymorphic organisms (Wigglesworth, 1954);, there are assumed the presence of convergent and divergent metamorphosis (Snodgrass, 1954, 1960, 1961);, there is analyzed the origin of pupa phase (Pojarkov, 1914; Hinton, 1963; etc.);, there are differed larvae into adaptive and phyletic (Cohen, Massey, 1983), not to speak about the definition of this phenomenon given in manuals (Grin at al., 1990) and encyclopedias.

From this, by no means a full list of available conceptions on metamorphosis, it is clear, however, that all its definitions, as special andso as more wide, mostly only describe from various positions a unique phenomenon of changing of the animal morphology during the life and do not reveal its sources and nature except for rather rare cases (Williamson, 1992). So, attempts to ascertain ontogenetic significance and value of this phenomenon are far from unnecessary means. It should be noted, that matter is not only of the terms but of the sense of this quite specific phenomenon, which is described as metamorphosis. Declaring high specificity of metamorphosis, as R. Snodgrass (1954) emphasized, we attach to it unique characters and, consequently, have to exclude from this consideration all other types of development.

Generally, metamorphosis as a change of the animal form, can be evolved nearly exclusively on the basis of more or less long ontogenesis, including a number of stages, when an animal has a time to adapt to different environments in the course of its ontogenesis, and "the exchange of habitat was a source of differences of adult insects from developing one" (Jezhikov, 1929, p. 14). With the same success but, certainly, on the basis of different ontogenic and morphological backgrounds, it would concern not only insects but also other animals. The initial stages of the so long ontogenesis could apparently appear in consequence of de-embryonization, i.e., hatching from the egg, owing to various reasons, on earlier stages of development regarding original ontogenesis that leads to formation of larva unlike to adult animal and forming various cenogenetic adaptations. It is possible that various characterizations and differentiations of larvae, for instance, on the basis of morphological features, functions (spreading, feeding), physiology or ontogenic criteria, which are found most convincing. That is way, the following definition of larva may be proposed: *a true larva is a particular stage (morph) of animal in the course of its individual development originated as a result of de-embryonization. , i.e., unlike adult animal, incapable of reproduction and developing various secondary cenogenetic adaptive characters, which have not been in adult precursors*. In other words, larva is a quite adapted to its habitat instar, and not underdeveloped as is frequently fought. Underdeveloped organs in larva are organs of adult animal, about which we, observing larva, cannot know anything because they remain in under-displayed conditions. It follows from this conception that larva as such (but not any un-reproducing form owing to accumulation of adaptive characters, as Shishkin (1988) supposed) exists only in the form of true larva, originated only owing to de-embryonization as a result of particular developmental reasons that, in differentiation of environmental conditions, leads to appearance of the secondary adaptive characters and, consequently, to metamorphosis in the aim to overcome differences between larva and adult animal. Thus, true larva is always differentiated.

Capability of such differentiated larva to independent evolution – is a quite debatable question, which, however, hasve been frequently resolved positively (Wigglesworth, 1954), especially taking into consideration the possibility of natural selection, i.e., selection of characters of intermediate developmental stages (Shishkin, 1988). This conception about own adaptive evolution of intermediate stages was known even to E. Haeckel, an author of bio-genetic rule, and also was many a times discussed of by his subsequent critics (see, for instance, Garstang, 1922). French acarolog F. Gandjean has gone even farther and grounded, in applying to mites, a conception termed "age-dependent evolution" (Grandjean, 1954, 1957, André, 1988), i.e., evolution dependent on the ontogenic age of a stage. Nevertheless, if remembering that evolution is realized as a selection, which is no one as differentiated reproducing of the most adapted organisms, statement of evolution of non-reproducing larval stage looks at least strange. Moreover, it is very problematic to consider acquisition of any cenogenetic characters by larva or any other stage as recognition of morphological evolution of this stage (instar) (Shichkin, 1988). In such a case, it has to inevitably differed differ in evolution of stages and organs from one hand and evolution of organisms proper in combination of all ontogenic phenomena, from other hand. As it was noted above, larva may acquire some new morphs only in the context of the aims of the whole ontogenesis, which evaluates always as a whole (Schmalghausen, 1982).

De-embryonization, which leads to formation of larva – is a transformation of the embryonic instar into post-embryonic one on some period of evolution of some animal group that could proceed in the early evolution in many if not all groups, including mites. Nevertheless, early hatching from the egg is not enough evidence for existence of true larva, i.e., its initial embryonic status that is seen on the example of mites. At present historical epoch, in the appearance of their larvae, there are not any embryonic characters, if leaveing out of consideration that all mites could originate from ancestors having larval status (Krivolutskiy, 1999) that is again seen from their internal morphology. Conversely, the origin of metamorphosis is a result of "earlier hatching of embryo from the egg that results that larva retains mostly embryonic organization" (Jezhikov, 1929, p. 25). Thus, if we'll manage to prove that larvae recapitulate morphological conditions peculiar to ancestral adult forms, this is embryonic instar. Larvae of mites, however, do not give us such opportunity. Six-legged condition of the mite's ancestors is quite doubtful. Nevertheless, larva of mites is unique, because it is a temporary feature of retardation of part of characters provoked by somatic condition of embryogenesis on the background of acceleration of some tissue systems. As a matter of fact, it is the first nymphal stage and, consequently, mites lack metamorphosis.

Coming back to differentiated true larva, it should be emphasized that its transformation to adult form is realized by means of true metamorphosis. Nevertheless, R. Snodgrass (1961) referred, as he supposed, true metamorphosis to temporary retreat of the wing insect to a condition of caterpillar (divergent metamorphosis), whereas its transformation to adult insect he termed as retromorphosis, i.e., return to initial reproducing form (convergent metamorphosis). Such interpretation seems to be relevant if assumeing that initially there was wing insect proper, and already afterwards, due to various ontogenetic processes, the caterpillar has been formed, but not contrary.

This temporary retreat in the sense of R. Snodgrass, as its source, – de-embryonization has probably been caused by necessity of elaboration of the mode of optimal feeding (trophism) for every given species (ontogenesis) in the purpose of the more effective realization of functions of mature organism, in particular, reproduction. As is known, insect larva, feeds and grows, in what is its main function. In many aquatic, in particular, see invertebrates, for instance, polychaetes an it is observed inverse satiation is observed – larvae realize function of spreading, whereas adult animals feed, i.e., reproduction is realized on the stage of growth and feeding. At the same time, they possess, as it is thought (Jezhikov, 1953), primary metamorphosis, because these spreading larvae are considered as initial. In this situation, there are obviously seen differences between primary and secondary, or true metamorphosis, when reproduction is realized on the stage of spreading, and, moreover, there are existed~~exist~~ true larvae engaged in feeding. Furthermore, there are quite different life strategies in these groups of animals (Highnam, 1981).

In other words, when growth, spreading and reproduction are focused within the same stage, metamorphosis is not necessary;, it is absent. Thus, metamorphosis needs when functions of spreading, feeding and reproduction are distributed in the ontogenesis among different stages. That is way, development with larva – widely adapted embryo – determines the presence of metamorphosis and so may be considered as special development. Un-special development – is the development with "'imaginary'" larva as, for instance, in mites, where these larvae do not possess embryonic characters and special functions. In general, however, larvae have obtained so many various self-dependent characters that they may have in the life of animal self-sufficient ontogenetic role, whereas existence of adult form (imagoe) is frequently very restricted by functions and time (Cohen, Massey, 1983).

Feeding of embryo in the egg thus appears un-effective~~ineffective~~ for providing of full-fledged adult (imaginal) life. So, larva and, consequently, metamorphosis, are possible and appear in ontogenesis in that time when the egg cannot be able to feed a developing organism up to needed conditions, i.e., when reserves of yolk in the egg are insufficient for a rapid achievement of adult condition by the organism (Schmalghausen, 1982). Such "'premature'" hatching proceeds even in that case when the egg is seem provided with a large amount of yolk that typically takes place in arthropods (Ivanova-Kazas, 1997), and in insects in most cases there is even no lecitotrophic larvae (Highnam, 1981).

Metamorphosis – is a result of divergence of developmental ways of larva and adult animal in every given ontogenesis, and transition between larval and mature (adult) programs in the model of their parallel development is metamorphos proper (Cohen, Massey, 1983). Especially, if larva, following these authors, possesses not only phyletic characters as, for instance, trochophora of polychaetes, but complex adaptive characters as larvae of holometabolous insects, i.e., in the case of secondary metamorphosis (Jezhikov, 1953). Thus, metamorphosis – is compression of development under a pressure of environment up to a minimum number of stages, ultimately up to one (pupa) in the case of morphological and ecological differences between larva and adult animal (Sehnal, 1985), i.e., difference in their developmental programs. In this case, larva in accordance with the theory of Berlesei-Jezhikov is considered as embryonic form undergone de-embryonization (Jezchikov,1929), and metamorphous organism would undergo rapid development (tahigenesis) of imaginal structures, including as somatic so generative elements. As concerns de-embryonization, in some cases it seems to be a direct result of acceleration, i.e., particular heterochronia, which is defined by a rapid rate of morphological development of some somatic characters (McNamara, 1986) leading to early hatching from the egg. And conversely, it is thought that embryonization in the case of the original long ontogenesis, is advantageous and even necessary for groups with similar mode of life of larva and imagoe (Tihomirova, 1974, 1991). In the case of the same habitat, larvae early or late would inevitably embryonized.

Partial or total immobility of pupa of holometabolous insects originated on the basis of this difference between larva and adult and lead to limitation of its communication with external milieu, provide further divergent adaptation of larva and adult form, combined nevertheless within the same history of life (Williamson, 1992). Thus, the pupa stage, always restricted in only one instar but having initially probably several nymphal instars (stages) (Jezhikov, 1929), is a progressive stage, which divide into more and more ecologically divergent larval and adult stages. On the question of the pupa origin, there are two polar points of view – theories of Berlesei-Imms-Jezhikov and Pojarkov-Snodgrass-Hinton. Not to stay on these theories in detail, it should be noted, however, that E.F. Pojarkov "'stretches'" ontogenesis up to flying imagoe with its differentiated and complex muscle system, in a consequence of which a pupa is needed as a matrix for its reorganization (Pojarkov, 1914; Hinton, 1963). Conversely, in accordance with I.I. Jezhikov (1929), appearance of the full transformation (transmutation) just leads to formation of imaginal muscles. These conceptions, however, do not take into account that initially there has probably been flying insect proper and only after that specialized larva, pupa and metamorphosis as such have evolved. Indeed, if larva would haves formed initially, it couldn't need to transform into flying form, and evolution would go on a quite different direction. Most likely, however, these processes evolved parallel in the historical retrospection.

From the general position, more number of stages as, for instance, in mites denies the idea of metamorphosis and significant morphological difference between larval and adult stages (instars) first of all on the character of internal organization including muscle system. So, observable difference between larva and imagoe, as in holometabolous insects, doubtfully is overcome first of all from ecological position, if all metamorphic processes would not be restricted by a single stage, namely pupa, relatively independent from external milieu and is not dependentd on feeding.

Embryonic status of true larvae, as I.I. Jezhikov (1929) supposed, defines an extremely rich development of their adaptive characters owing to adaptation to habitat different from the habitat of adult animal. Hypothetical transfer of reproduction function to larval stage (progenesis) (Cohen, Massey, 1983) or hybridization (Williamson, 1992) would give a great explosion of species formation and would lead to a large amount of essential new forms (Garstang, 1922). In other words, if larvae could receive a possibility of independent evolutionary development, sufficiently new groups of the level of class could form (Jezhikov, 1929).

It may be asserted, therefore, that metamorphosis as such – is a quite specific phenomenon characterized, first of all, by distinct criteria of ontogenetic character. In accordance with the above -mentioned ideas, it may be proposed the following definition of metamorphosis may be proposed. *Metamorphosis – is a rather sudden, inconvertible and strong change of morphology and mode of life of a specimen during its individual development and related to transfer of reproduction function from one differentiated form (larva) to other (adult animal) in condition that both of these forms are adapted to different environmental conditions.* In all other cases, development wouldn't satisfy theto specific character and sense of metamorphosis, i.e., it couldn't be attributed as true metamorphosis.

CONCLUSION

In this article, atn attempt is made to clarify only some problems of arthropod ontogenesis most closest to the author's interest. Many questions remain beyond discussion, first of all concerning genetic of ontogenesis. At the same time, it should be noted that in consideration of different evolution trends in animal kingdom, for unknown reason, for the first time, it is given preference is given for the first time to neotenie (Ivanova-Kazas, 1997 b, c), although the equal significance, if not more, has such a phenomenon as progenesis (Smirnov, 1991). In particular, progenesis, reproduction of true larvae, could and probably has lead to formation of new forms and even animal groups (for instance, Acari), and not in the least neotenie – retaining of some larval characters in sexual mature organism on the a stage equivalent to adult organism of historical previous ontogenesis. Just in the case of neotenie, metamorphosis is lacking from ontogenesis because in such mode of development larval characters are not overcome but retain in ontogenesis.

In the broad sense, there is no metamorphosis without larva. Larva – is always immature but not an un-differentiated form. Larva may have many functions – feeding, spreading, even multiplication – but never direct forms of sexual reproduction. Otherwise, it is either not a larva or a deep aberrant form. One's nature of larva does not allow it to reproduce. It appears from the egg, just as a result of insufficient quantity of yolk and just for the aim to achieve rapidly, owing to intensive feeding, sexually mature condition of organism. This phenomenon, which may be classified as de-embryonization, leads in the subsequent evolution to metamorphosis, i.e., special development, including two different forms, one of which bears the function of reproduction.

ACKNOWLEDGMENTS

This study is realized owing to support from the Russian Foundation for Fundamental Research by the Grant N 09-04-00390-a.

REFERENCES

Anderson, D.T. (1973). Embryology and phylogeny in *Annelids and Arthropods*. Oxford: Pergamon Press.

André, H.M. (1988). Age-dependent evolution: from theory to practice. *Ontogeny and Systematics*. New York: Columbia University Press.

Bei-Bienko, G.Ja. (1980). General entomology. *Moscow: Vischaya shkola*. (In Russian).

Bergström, J. (1986). Metazoan evolution – a new model. *Zool. Scr.*, 15: 189-200.

Bergström, J. (1992). The oldest arthropods and the origin of the Crustacea. *Acta Zoologica* (Stockholm), 73: 287-291.

Canard, A. & Stockmann, R. (1993). Comparative postembrionicpostembryonic development of Arachnids. *Proc. XII Int. Congr. Arachnol. Mem. Queensl. Mus.*, 33: 461-468.

Cisne, J.L. (1974). Trilobites and the origin of arthropods. *Science*, 186: 13-18.

Cohen, J. & Massey, B.D. (1983). Larvae and the origins of major phyla. *Biol. J. Linn. Soc.*, 19: 321-328.

Cutler, B. (1980). Arthropod cuticle feature and arthropod monophyly. *Experientia*, 36: 953.

Dubinin, V.B. (1959). Cheliceronous animals (subtype Chelicerophora W. Dubinin nom.n.) and their position in the system. *Zool. Zhurn.*, 38, 1163-1189. (In Russian).

Fryer, G. (1996). Reflections on arthropod evolution. *Biol. J. Linn. Soc.*, 58: 1-55.

Garstang, W. (1922). The theory of recapitulation: A critical re-statement of the biogenetic law. *J. Linn. Soc. Lond.* (Zoology), 35: 81-101.

Grandjean, F. (1954). Les deux sortes de temps et l'evolution. *Bull. Biol. France Belg.*, 88 : 413-434.

Grandjean, F. (1957). L'evolution selon l'age. *Archs. Sci. Phys. Natur. Geneve*, 10: 477-526.

Grin, N., Staut, W., Teilor D. (1990). *Biology.* Vol. 3. Moscow: Mir. (Translation from English).

Highnam, K.C. (1981). A survey of invertabrateinvertebrate metamorphosis. *Metamorphosis. A problem in developmental biology.* New York-London: Plenum Press.

Hinton, H.E. (1963). The origin and function of the pupal stage. *Proc. Roy. Ent. Soc. London, Ser. A*, 38: 77-85.

Ivanov, P.P. (1937). General and comparative embryology. Moscow-Leningrad: *Biomedgiz.* (In Russian).

Ivanova-Kazas, O.M. (1979). Comparative embryology of invertebrate animals. *Arthropods.* Moscow: Nauka. (In Russian).

Ivanova-Kazas, O.M. (1997 a). Strategy, tactics and evolution of ontogenesis. *Ontogenes.* 28: 31-40.

Ivanova-Kazas, O.M. (1997 b). Neotenie as particular mode of evolution. 1. Neotenie in lower Metazoa, polychaetes amd and mollusks. *Zool. Zhurn.*, 76: 1244-1255. (In Russian).

Ivanova-Kazas, O.M. (1997 c). Neotenie as particular mode of evolution. 2. Neotenie in arthropods and chordates. *Zool. Zhurn.*, 76: 1256-1265. (In Russian).

Ivanova-Kazas, O.M. & Ivanov, A.V. (1987). On the theory of "'trochea'" and on phelogenetic significance of ciliary larvae. *Biol. Morja*, 2: 6-21. (In Russian).

Jezhikov, I.I. (1929). Metamorphosis of insects. Moscow: *Publisher of Moscow State University.* (In Russian).

Jezhikov, I.I. (1953). Characteristics of the early embryonal stages in total and un-total transformation of insects. Proc. Int. Morph. Anim. in the name of A.N. *Severtsov* . 8: 130-153. (In Russian).

Kamshilov, M.M. (1970). Organization and evolution. *Zhurn. Obsch. Biol.*, 31: 157-178. (In Russian).

Kamshilov, M.M. (1972). Ontogenesis and evolution. Regularity of the progressive evolution. Leningrad: *Publishers of Inst. Hist. Estestv. Techn.*: 168-185. (In Russian).

Kluge, A.C. (1988). The characterization of ontogeny. *Ontogeny and Systematics.* New York: Columbia University Press: 57-81.

Krivolutskiy, D.A. (1999). Phylogeny and taxonomical rang of mites. VII Acarological Conference. Theses of Reports. St.- Petersburg: 36-37. (In Russian).

Locke, M. (1981). Cell structure during insect metamorphosis. *Metamorphosis. A problem in developmental biology.* New -York-London: Plenum Press, 75-103.

Locke, M. (1985). A structural analysis of post-embryonic development. *Comp. Insect Physiol. Biochem. Pharmacol.* Oxford, etc.: Pergamon Press, 2: 87-149.

Manton, S.M. (1973). Arthropod phylogeny – a modern synthesis. *J. Zool.*, London, 171: 111-130.

McNamara, K.J. (1986). A guide to the nomenclature of heterochrony. *J. Paleontol.*, 60: 4-13.

Minichev, Ju.S. (1997). Endolarval and plagiaxonal types of developments of Polychaete. *Zool. Zhurn.*, 76: 1300-1307. (In Russian).

Nielsen, C. (1985). Animal phylogeny in the light of the trochaea theory. *Biol. J. Linn. Soc.*, 25: 243-299.

Nielsen, C. (1994). Larval and adult characters in animal phylogeny. *Amer. Zoologist*, 34: 492-501.

Nielsen, C. (1997). The phylogenetic position of the Arthropoda. *Arthropod Relationships*. London: Chapman & Hall, 11-22.

Nielsen, C. (1998). Origin and evolution of animal life cycles. *Biol. Rev.*, 73: 125-155.

Otto, J.C. (1997). Observations on prelarvae in Anystidae and Teneriffiidae. Acarology IX, Proceedings. Columbus: *Ohio Biol. Survey*, 343-354.

Pojarkov, E.F. (1914). Essay on the theory of pupa of insects with total transformation. *Proc. Russian Ent. Soc.* 41: 1-51. (In Russian).

Queiroz, K. de. (1985). The ontogenetic method for determining character polarity and its relevance to phylogenetic systematics. *Syst. Zool.*, 34: 280-299.

Sehnal, F. (1985). Growth and life cycle. *Comp. Insect Physiol. Biochem. Pharmacol.* Oxford, etc.: Pergamon Press, 2: 2-86.

Severtsov, A.S. (1970). On the question of evolution of ontogenesis. *Zhurn. Obsch. Biol.* 31: 222-235. (In Russian).

Sharov, A.G. (1965 a). Origin and main stages of arthropod evolution. 1. From annelids to arthropods. *Zool. Zhurn.*, 44: 803-817. (In Russian).

Sharov, A.G. (1965 b). Origin and main stages of arthropod evolution. 2. Origin and phylogenetic relationships of the main arthropod groups. *Zool. Zhurn.*, 44: 963-979. (In Russian).

Shatrov, A.B. (1991). Some problems of ontogenesis of trombidiform mites (Acariformes, Actinedida). *Parazitology*, 25: 377-387. (In Russian).

Shatrov, A.B. (1993). Individual development and characteristics of its differentiation in arthrrropods. *Entomol. Rev.*, 72: 441-455. (In Russian).

Shatrov, A.B. (1998). Ecologo-physiological analysis of ontogenesis in parasitengona mites. *Parasitology*, 32: 385-395. (In Russian).

Shatrov, A.B. (1999). Contribution to the prelarva status: the moultingmolting cycle of the calyptostasic prelarva of the trombiculid mite Leptotrombidium orientale (Acariformes: Trombiculidae). *Acarologia*, 40: 265-274.

Shatrov, A.B. (2000). Trombiculid mites and their parasitism on vertebrate animals. St. .-Petersburg: Publishers of St. -Petersburg University. (In Russian with large English summary).

Shishkin, M.A. (1988). Regularity of ontogenesis evolution. *Modern Paleontology*. Moscow, Nedra, 2: 169-209. (In Russian).

Shmalghausen, I.I. (1982). Organism as a whole in individual and historic development. *Collected works*. Moscow: Nauka. (In Russian).

Smirnov, S.V. (1991). Pedomorphosis as a mechanism of evolutionary transformations of organisms. *Modern evolutionary morphology*. Kiev: Nauk. Dumka: 88-103. (In Russian).

Snodgrass, R.E. (1954). Insect metamorphosis. Smithson. *Misc. Coll.*, 122: 1-124.

Snodgrass, R.E. (1960). Some words and their ways in entomology. *Proc. Ent. Soc. Wash.*, 62: 265-270.

Snodgrass, R.E. (1961). Insect metamorphosis and retramorphosis. *Trans. Amer. Entomol. Soc.*, 87: 273-280.

Starobogatov, Ja.I. (1991). Phylogeny and system of arthropods. *Progr. Mod. Biol.*, 111: 828-839. (In Russian).

Tichomorova, A.L. (1974). De-embryonization as mechanism of phylogenetic transformations in order Coleoptera. *Zhurn. Obsch. Biol.*, 35: 620-630. (In Russian).

Tichomirova, A.L. (1991). Reconstruction of ontogenesis as mechanism of insect evolution. Moscow: Nauka. (In Russian).

Vainstein, B.A. (1978). Identification keys of the soil inhabiting mites. *Trombidiformes.* Moscow: Nauka. (In Russian).

Wald, G. (1981). Metamorphosis: an overview. *Metamorphosis.* A problem in developmental biology. New York-London: Plenum Press, 1-39.

Weygoldt, P. (1986). Arthropod interrelationships – the phylogenetic-systematic approach. *Z. Zool. Syst. Evolut.*, 24: 19-35.

Wigglesworth, V.B. (1954). The physiology of insect metamorphosis. Cambridge: Cambridge University Press.

Wilbur, H.M. (1980). Complex life cycles. *Ann. Rev. Ecol. Syst.*, 11: 67-93.

Williamson, D.J. (1992). Larvae and evolution: *Towards a new zoology.* New York: Chapman & Hall.

Wolpert, L. (1990). The evolution of development. *Biol. J. Linn. Soc.*, 39: 109-124.

Zachvatkin, A.A. (1953). Investigation on the morphology and post-embryonic development of tyrogliphids (Sarcoptiformes, Tyroglyphoidea): Collected scientific works. Moscow: Publishers of Moscow University: 19-120. (In Russian).

Zachvatkin, A.A. (1975). Embryology of Insects. Moscow: *Vischaya shkola.* (In Russian).

In: Larvae: Morphology, Biology and Life Cycle
Editors: Kia Pourali and Vafa Niroomand Raad

ISBN: 978-1-61942-662-7
© 2012 Nova Science Publishers, Inc.

Chapter 3

EFFECTS OF PLANT LECTINS AND TRYPSIN INHIBITORS ON DEVELOPMENT, MORPHOLOGY AND BIOCHEMISTRY OF INSECT LARVAE

Patrícia M. G. Paiva, Emmanuel V. Pontual, Thiago H. Napoleão and Luana C. B. B. Coelho*

Departamento de Bioquímica, Centro de Ciências Biológicas, Universidade Federal de
Pernambuco, Cidade Universitária, Recife, Pernambuco, Brazil

ABSTRACT

Lectins are hemagglutinating proteins that promote cellular responses due to their interaction with glycosylated molecules. Trypsin inhibitors have been described as endogenous regulators of proteolytic enzymes and as storage proteins. Plant lectins and trypsin inhibitors with insecticidal activity may become an alternative to synthetic insecticides that adversely affect the environment and have promoted the emergence of resistant organisms. These proteins exert deleterious effects on larval survival, weight, feeding ability, and development of insects as well as morphology of larvae, pupae and adults. The digestive processes in larval gut are highly active and their integrity is an essential aspect in insect development. Plant lectins and trypsin inhibitors can modulate the activity of digestive enzymes in larval gut. This chapter reports larvicidal lectins and trypsin inhibitors isolated from different plant tissues (e.g. bark, heartwood, leaves and seeds) with potential applications in the control of numerous insect species, including *Achaea janata*, *Aedes aegypti*, *Anthonomus grandis*, *Callosobruchus maculatus*, *Corcyra cephalonica*, *Ephestia kuehniella*, *Helicoverpa armigera*, *Lacanobia oleracea*, *Ostrinia nubilalis*, *Spodoptera littoralis*, *Spodoptera litura* and *Zabrotes subfasciatus*. The changes in morphology of larvae treated with lectins or trypsin inhibitors, as well as the effect of these proteins on larval α-amylase, β-glucosidase, protease and aminopeptidase activities, are also discussed.

---

* Corresponding author. Tel: +558121268540; fax: +558121268576. E-mail: ppaivaufpe@yahoo.com.br.

1. PLANT LECTINS:
STRUCTURE AND BIOLOGICAL ACTIVITIES

Lectins, proteins or glycoproteins non-immune in origin, interact with carbohydrates through binding sites that generally are shallow depressions located on the protein surface (Ambrosi et al., 2005). The linkage between lectins and carbohydrates occurs non-covalently, and involves hydrogen bonds as well as van der Waals and hydrophobic interactions.

Barks, flowers, leaves, roots, and seeds are sources of lectins with high structural and functional diversity (Coelho et al., 2009; Sá et al., 2009b; Santos et al., 2009; Napoleão et al., 2011; Souza et al., 2011). Regarding molecular arrangement, lectins may vary considerably, from monomeric proteins to proteins presenting different subunits per molecule (Bhat et al., 2010; Yan et al., 2010). Intermolecular forces involved in association of lectin subunits are hydrophobic interactions, disulfide bridges and hydrogen bonds (Kennedy et al., 1995). Plant lectins are involved in cellular processes, including modulation of enzyme activities, mediation of symbiotic rhizobia bacteria attachment to roots, and protection against insect and phytopathogenic microorganisms (Rüdiger and Gabius, 2001; Rodriguez-Navarro et al., 2007; Sá et al., 2008; Michiels et al., 2010).

Lectins agglutinate cells and precipitate polysaccharides, glycoproteins and glycolipids, though with no structural modification. The binding of lectins to glycoconjugates on the surface of erythrocyte forms a network that maintains the erythrocytes agglutinated and suspended in solution, in what is called hemagglutinating activity. This biological property can be detected using microtiter plates (Figure 1A). Plant tissues contain compounds that can disperse erythrocytes, promoting the typical hemagglutination aspect observed in microtiter plates (Correia et al., 2008).

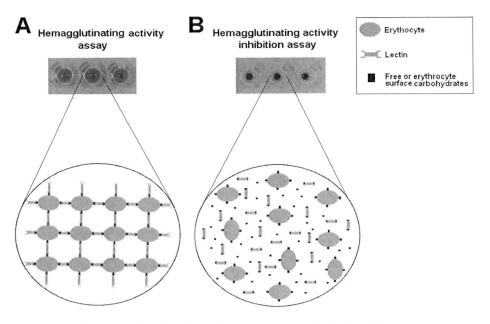

Figure 1. Aspects of hemagglutinating (A) and inhibitory hemagglutinating (B) assays in microtiter plates. The circles show schematic representations of the erythrocyte network formed by lectin (A) and inhibition of lectin activity by free carbohydrate molecules (B).

To ensure that erythrocytes are held in suspension due to agglutinating effect (lectin activity), crude samples should be evaluated using an inhibitory assay of hemagglutinating activity with carbohydrate or glycoprotein preparations (Figure 1B). The inhibitory assay, in addition, establishes the lectin specificity for carbohydrates; according to this criterion, lectins can be classified into fucose, mannose, sialic acid, N-acetylglucosamine, N-acetylgalactosamine and glycan-complex groups (Peumans and Van Damme, 1998).

Purified lectins have been used for arrays of diagnosis purposes based on the interaction with cell glycoconjugates. It has been established that the lectin microarray technique differentiates closely related bacteria strains and reveals differences in the glycan structure of the alpha-fetoprotein – a serum marker of liver cancer – from human hepatocarcinoma and cord serum (Chen et al., 2008; Gemeiner et al., 2009).

Plant lectins present several interesting biological properties. Seed lectin from *Phaseolus vulgaris* was active against HIV-1 reverse transcriptase and inhibited the proliferation of human tumor cells by inducing the production of apoptotic bodies (Fang et al., 2010). Lectin from *Cratylia mollis* seeds (Cramoll 1,4) inhibited the proliferation of *Trypanosoma cruzi* epimastigotes and was effective in tissue repair in mice experimentally challenged with injury (Fernandes et al., 2010; Melo et al., 2011). Lectins from *Bauhinia monandra* secondary roots and *Myracrodruon urundeuva* heartwood have been shown to be antifungal agents against phytopathogenic species of *Fusarium* (Sá et al., 2009a; Souza et al., 2011). Lectins from *Eugenia uniflora* seeds, *M. urundeuva* heartwood and *Moringa oleifera* seeds showed antibacterial activity against *Staphylococcus aureus* and minimal inhibitory concentrations were 1.5, 0.58, and 7.8 µg/ml, respectively. These values corresponded to the concentrations of those clinically relevant anti-staphylococcal agents (Gibbons, 2004; Oliveira et al., 2008; Sá et al., 2009a; Ferreira et al., 2011). The activity of lectins against insect larvae will be discussed later.

2. PLANT TRYPSIN INHIBITORS: STRUCTURE AND BIOLOGICAL ACTIVITIES

Plants produce proteins with protease inhibitor activity in an attempt to control the activity of self proteases (Laskowski Jr. and Qasim, 2000). The inhibitor-enzyme interaction results in stable complexes that are either inactive or present low enzyme activity (Liao et al., 2007). The synthesis of protease inhibitors by plants can be induced by water stress, seasonal variations in temperature or luminosity and, especially in response to infection by pathogens (Bray et al., 2000; Kuk et al., 2003; Bhattacharyya et al., 2007).

Proteinaceous trypsin inhibitors bind tightly and competitively to the enzyme active site, preventing the interaction with the substrate (Bode and Huber, 2000). The enzyme/inhibitor interaction occurs through the establishment of hydrophobic and electrostatic interactions as well as hydrogen bonds between the active site of the protease and the reactive site in the inhibitor molecule. The reactive site may be located on a protruding active loop and enzyme inhibition occurs when the amino acid sequence in the inhibitor reactive site is complementary to that in the enzyme active site (Major and Constabel, 2007).

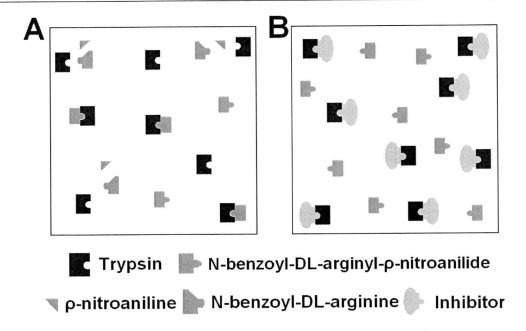

Figure 2. Scheme showing the hydrolysis of N-benzoyl-DL-arginyl-ρ-nitroanilide (BApNA) by trypsin (A) and inhibition of the enzyme activity by complexing with a competitive proteinaceous inhibitor (B).

Mutations in the reactive site may lead to loss of inhibition ability, as reported for the chymotrypsin/trypsin inhibitor isolated from *Psophocarpus tetragonolobus* seeds (Roy et al., 2011). Despite the high specificity of inhibitors, some may interact with more than one enzyme due to the presence of two independent reactive sites that bind to different enzymes; an example is the *Adenanthera pavonina* Kunitz trypsin inhibitor that inhibits simultaneously trypsin and papain (Migliolo et al., 2010). Proteinaceous trypsin inhibitors have demonstrated several biological applications such as antiviral, antifungal, antibacterial, and antitumoral activities.

Trypsin hydrolyses peptide bonds preferentially in arginyl residues, and the chromogenic substrate N-benzoyl-DL-arginyl-ρ-nitroanilide (BApNA) is used in trypsin activity assays. BApNA hydrolysis releases ρ-nitroaniline (Figure 2A), which absorbs light at wavelengths in the yellow region of the spectrum, in the reaction medium. The presence of a trypsin inhibitor in the sample can be detected as the absence of ρ-nitroaniline in the reaction medium (Figure 2B).

3. INSECT PESTS AND VECTORS LARVAE

Insect larvae naturally attack crops and stored plant products, causing serious damage to agriculture and economy. Larvae of *Callosobruchus maculatus* (cowpea weevil) and *Zabrotes subfasciatus* (Mexican bean weevil) severely affect seed quality and storability, larvae of *Corcyra cephalonica* (rice moth) and *Ephestia* (*Anagasta*) *kuehniella* (Mediterranean flour moth) are pests of stored flour and grain products, the larvae of *Lacanobia oleracea* (bright-line brown-eye) attack tomato crops, and larvae of *Ostrinia nubilalis* (European corn borer) infest a variety of grains, particularly maize (Blackford and Dinan, 1997; Allotey and Azalekor, 2000; Macedo et al., 2007).

Larvae of *Achaea janata* (castor semilooper) are pests of hoop pine (*Araucaria cunninghamiis*) seedlings, tamarind (*Tamarindus indica*) and castor oil plants (*Ricinus communis*). Additionally, insect larvae can cause injuries to aerial parts of plants, impairing survival or reproduction. *Anthonomus grandis* (cotton boll weevil) and *Helicoverpa armigera* (cotton bollworm) affect flowering and fruiting structures of many economically important crops, while *Spodoptera litura* (cutworm) can defoliate apple orchards (*Malus domestica*) as well as, banana (*Musa paradisiaca*), coffee (*Coffea* sp.), corn (*Zea mays*), cotton (*Gossypium* sp.) and tobacco (*Nicotiana tabacum*) plantations (Franco et al., 2003; Jallow et al., 2004; Bhattacharyya et al., 2007). *Spodoptera littoralis* (Egyptian cotton leaf-worm) is a major polyphagous pest, common in Africa, Mediterranean Europe, and Asia; at least 87 economically important plant species are targeted by these caterpillars, which feed on their leaves, flowers and fruits (Azab et al., 2001).

Insects are also vectors of viral diseases that threaten human health. Infected *Aedes aegypti* females can transmit yellow fever, dengue fever, which is an acute hemorrhagic infection that causes a severe flu-like illness, and chikungunya fever, a disease that causes severe joint pain and has no cure (World Health Organization, 2008; World Health Organization, 2009a,b). The control of mosquito populations is fundamental to reduce spreading of diseases; in this sense, new synthetic and natural insecticides have become the object of considerable research in recent times.

Insecticidal compounds can affect indiscriminately all insect life stages, or act against mature or immature forms preferentially. The control of an insect population by targeting only adults may not be a successful strategy, since immature forms will persist, proceeding with the life cycle of the species in question. On the other hand, compounds with larvicidal effect can be more effective, as they prevent the emergence of adult forms of vectors.

Chemical insecticides, such as organophosphorus, carbamates, organochlorine and pyrethroids have been widely employed to control larvae of disease vectors and pests. Organophosphorus and carbamates inhibit acetylcholinesterase activity, while organochlorine and pyrethroids keep the sodium channels of neuronal membrane open (Forsberg and Puu, 1984; Coats, 1990). The downside of these pesticides is that they are not selective, affecting both the insect populations and non-target organisms. For this reason, when indiscriminately used these compounds may have adverse effects on human health, including reduced fertility and fecundity, abnormalities of reproductive tracts, neurobehavioural disorders, and several types of cancer (McKinlay et al., 2008). Moreover, the continuous use of the chemical pesticides with no rotation planning has induced the selection of resistant individuals, eventually leading to increasing lethal concentrations required to kill 50% of larvae (LC50) and to the development of resistant populations (Hemingway and Ranson, 2000; Nauen, 2007). In this scenario, the search for new effective and biodegradable insecticides has become appealing. Lectins and proteinaceous trypsin inhibitors have been evaluated as alternatives to synthetic insecticides currently used to control insect larvae. The sources of larvicidal lectins and trypsin inhibitors discussed in this chapter as well as the insects affected by these proteins are listed in Table 1.

Table 1. Sources of plant insecticidal proteins and insect species affected by them

Scientific name	Common name	Affected insect
Lectin source		
Allium sativum	Garlic	*Dysdercus cingulatus, Lypaphis erysimi, Spodoptera littoralis*
Annona coriacea	Araticum (Portuguese)	*Corcyra cephalonica, Ephestia kuehniella*
Arachis hypogaea	Peanut	*Callosobruchus maculatus*
Bauhinia monandra	Butterfly tree, Napoleon's plume, pulse	*C. maculatus, E. kuehniella, Zabrotes subfasciatus*
Canavalia ensiformis	Jackbean	*Acyrthosiphon pisum*
Datura stramonium	Jimson weed	*C. maculatus*
Galanthus nivalis	Common snowdrop	*Lacanobia oleracea*
Griffonia simplicifolia	Griffonia	*C. maculatus*
Koelreuteria paniculata	Goldenrain tree	*E. kuehniella*
Maclura pomifera	Osage orange	*C. maculatus*
Moringa oleifera	Horseradish tree	*Aedes aegypti*
Myracrodruon urundeuva	Aroeira (Portuguese), urundel (Spanish)	*A. aegypti*
Solanum tuberosum	Potato	*C. maculatus*
Talisia esculenta	Pitomba (Portuguese)	*C. maculatus*
Triticum vulgaris	Wheat germ	*C. maculatus, Drosophila melanogaster, Ostrinia nubilalis*
Proteinaceous trypsin inhibitor source		
Adenanthera pavonina	Coralwood, Red sandalwood	*C. maculatus*
Archidendron ellipticum	Bangkong (Javanese)	*Spodoptera litura*
Cajanus cajan	Red gram	*Achaea janata*
Cicer arietinum	Chick pea	*Anthonomus grandis*
Crotalaria pallida	Smooth rattlebox	*C. maculatus*
Ipomoea batatas	Sweet potato	*S. litura*
Momordica charantia	Bitter melon	*Helicoverpa armigera*
Plathymenia foliolosa	Vinhático	*E. kuehniella*

4. EFFECTS OF PLANT LECTINS ON LARVAL DEVELOPMENT

Lectins act against immature and mature forms of insects by promoting deleterious effects on nutrient digestion and absorption. In general, insecticide lectins are resistant to proteolysis by insect digestive enzymes (Macedo et al., 2007; Napoleão et al., 2011, 2012; Oliveira et al., 2011). The insecticidal mechanisms can involve lectin binding to glycosylated molecules present in gut lumen, epithelial cells and peritrophic matrix (Macedo et al., 2007; Napoleão et al., 2012). Recently, it was reported that the binding of lectins to glycoproteins forming the membrane of midgut epithelial cells can start a signaling transduction cascade which leads to cell death (Hamshou et al., 2010). Figure 3 shows these potential lectin targets in larva midgut. Lectins can also enter the larva body by transcytosis across the midgut epithelium, interacting with hemocytes in the hemolymph, and reaching Malpighian tubules, ovaries and fat tissue (Fitches et al., 2001). The importance of the carbohydrate binding site for insecticidal activity of lectins has been demonstrated. In a study on the recombinant *Griffonia simplicifolia* leaf lectin II (rGSII) and mutant forms of rGSII with high, low or no *N*-acetylglucosamine-binding activity, Zhu-Salzman et al. (1998) showed that the insecticidal effect of this protein depends on the integrity of its carbohydrate binding site.

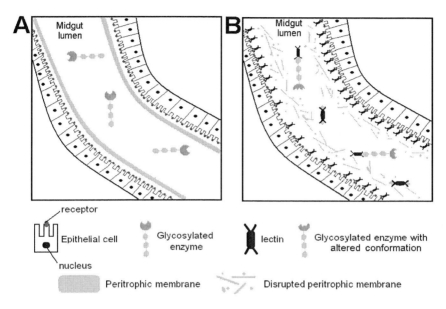

Figure 3. (A) Schematic representation of insect larva midgut showing epithelial cells, peritrophic membrane and glycosylated enzymes. (B) Larvicidal activity of lectins may be due to structural disorganization and disruption of peritrophic matrix integrity, modulation of enzyme activity and binding to receptors present in epithelial cells.

The rGSII variants that retained lectin activity significantly prolonged the development time of *Callosobruchus maculatus* larvae when incorporated into artificial seeds. The mutant protein with low *N*-acetylglucosamine-binding ability exhibited intermediate insecticidal activity; forms that were not able to recognize carbohydrates were not insecticidal.

Inhibitory effects on development have been described in larvae reared on artificial diets containing plant lectins as well as larvae fed on transgenic tissues expressing lectins. The delay in development is generally associated with impairment of larval growth. The binding of lectins to glycosylated proteins in the midgut of larvae inhibits nutrient uptake and reduces efficiency of diet utilization, causing a drop in mass gain. In general, the metamorphosis to pupa is strongly jeopardized and cannot occur.

The lectin from *Bauhinia monandra* leaf (BmoLL) blocked the development of *C. maculatus* larvae by reducing the number of surviving larvae and decreasing body weight. The lethal dose of BmoLL to kill 50% of larvae was 0.32%. The survival of *Zabrotes subfasciatus* larvae was also affected by BmoLL (LD_{50} of 0.5%), which also induced a 20%

decrease in body weight of larvae. Incubation of BmoLL with *C. maculatus* and *Z. subfasciatus* midgut extracts revealed that larvicidal effects may be linked to resistance to proteolysis by larval enzymes. The BmoLL-Sepharose matrix, in addition, bound midgut proteins from midgut cell membranes of *C. maculatus* larvae. Otherwise, supplementation of artificial diet with 1% BmoLL did not significantly decrease *Ephestia kuehniella* survival, though it caused a 40% decrease in larval weight. The subunits of BmoLL also exhibited resistance to digestion by proteolytic enzymes from midgut of *E. kuehniella* larvae (Macedo et al., 2007).

Talisia esculenta seed lectin (TEL) showed larvicidal activity against *C. maculatus* and the lethal dose was approximately 1% (w/w). Cysteine proteases from larva midgut were not able to hydrolyze TEL and the matrix TEL-Sepharose bound proteins from midgut of *C. maculatus* larvae (Macedo et al., 2004a). The lectins from *Maclura pomifera* (osage orange), *Arachis hypogaea* (peanut), *Solanum tuberosum* (potato), *Datura stramonium* (jimson weed) and *Triticum vulgaris* (wheat germ) also deleteriously affected *C. maculatus* larvae (Murdock et al., 1990). Wheat germ agglutinin (WGA) was also active against the third-instar *Ostrinia nubilalis* larvae, with a significant decrease in body mass gain of larvae fed on diet containing WGA (1.3 times), when compared to the control larvae (3.9 times), 24 h after ecdysis (Harper et al, 1998).

Evaluations of the effects of lectins on *E. kuehniella* larvae showed that the lectin from seeds of *Koelreuteria paniculata* (KpLec) was able to kill (LD_{50} of 0.65%) and reduce body weight (ED_{50} of 0.2%), while the coagulant lectin from *Moringa oleifera* seeds (cMoL) induced nutritional disturbances and delay in development of larvae as well as reduction in weight and survival of pupae (Macedo et al., 2003; Oliveira et al., 2011).

The *Annona coriacea* seed lectin (ACLEC) showed insecticidal activity against *E. kuehniella* larvae causing mortality (LD_{50} of 1.5%) and reducing body weight (ED_{50} of 1.0%). ACLEC was resistant to proteolysis by larval midgut enzymes for 48 h. On the other hand, ACLEC was not able to decrease the survival and mass of *Corcyra cephalonica*. However, two proteins from *C. cephalonica* larval gut extract adsorbed on an ACLEC-Sepharose column indicated that this lectin remained active in the larval gut, although it did not kill the larvae. Despite the ability of lectins to negatively affect larva survival or development, in the course of evolution insect pests efficiently evolve strategies to bypass or destroy the molecules of toxic compounds in their diets, adapting to these agents.

Interestingly, ingestion of ACLEC increased the efficiency with which *C. cephalonica* larvae converted the digested food into biomass, decreasing metabolic cost. *C. cephalonica* larvae fed on ACLEC showed decreased food consumption and faecal production. Such findings point to a compensatory effect, which reflects the ability of some lepidopterans to maintain growth rate and body mass, independently of food quality, by altering food consumption and use (Coelho et al., 2007).

Insecticide lectins have been incorporated into the genome of plants susceptible to insect attack and their protective effect on transgenic plants has been evaluated in laboratory or glasshouse conditions.

The effect of *Galanthus nivalis* lectin (GNA) on *Lacanobia oleracea* was evaluated by feeding the larvae with: I) an artificial diet containing GNA at 2% (w/v), II) excised leaves from transgenic potato plants expressing GNA at approximately 0.07% of total soluble proteins, and III) transgenic potato plants expressing GNA at approximately 0.6% of total soluble proteins. This assay was conducted in a glasshouse. GNA killed larvae (60% survival rate) in bioassay III, and reduction of larvae biomass of 32%, 23% and 48% was detected after feeding on diets I, II and III, respectively.

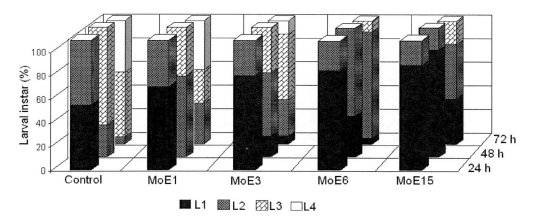

Figure 4. *A. aegypti* larval instars (%) after incubation of newly-hatched larvae for 24, 48, and 72 h with distilled water (control) and *M. oleifera* seed extracts (MoE1, MoE3, MoE6, and MoE15).

The ingestion of GNA by larvae also promoted delay in development, by approximately 2 days for each stage (diet I), and by 1.3 days for fourth-instar larvae (diet II). The study also showed that the delay in development was counterbalanced by shorter or similar durations of fourth, fifth and sixth instars in comparison with control group (Fitches et al., 1997).

Bandyopadhyay et al. (2001) reported that *Allium sativum* leaf lectin binds to glycosylated portions of specific receptors in the brush border membrane from the midgut epithelium of *Lypaphis erysimi* (mustard aphid) and *Dysdercus cingulatus* (red cotton bug) nymphs. The *A. sativum* lectin was also an insecticidal agent against *Spodoptera littoralis* (Sadeghi et al., 2008). The larvae fed on transgenic tobacco plants expressing this lectin showed reduced survival and weight gain; also, the size and weight of pupae and adults were lower than those of the control group.

The indiscriminate use of insecticides to control *Aedes aegypti* populations has induced the development of resistant larvae. Thus, lectins have been evaluated as promising alternatives to these insecticides.

Coelho et al. (2009) described the effect of aqueous extracts prepared with one, three, six, or fifteen *Moringa oleifera* seeds (MoE1, MoE3, MoE6 and MoE15, respectively) on the development of first-instar (L1) larvae of *A. aegypti* (Figure 4). After the first 24 h, 50% of larvae incubated in distilled water (control) had reached the second instar (L2), while in all extract treatments the first instar (L1) was predominant. After 48 h, larvae from control were predominantly at third instar (L3), which was not observed in MoE6 and MoE15 treatments at this experimental time. After 72 h, fourth-instar larvae (L4) were found in the control and MoE1, and to a lesser extent in MoE3 treatment. Larval development in MoE6 or MoE15 ceased in L3. The extracts from *M. oleifera* seeds showed lectin activity; the authors suggested that this class of insecticidal protein was involved in the delay in larval development promoted by extracts. Evaluation of water-soluble lectin (WSMoL) isolated from seeds for larvicidal activity showed that WSMoL promoted L4 mortality, with LC_{50} of 0.197 mg/mL (Coelho et al., 2009).

Three chitin-binding lectins with larvicidal activity against *Aedes aegypti* were isolated from *Myracrodruon urundeuva* bark, heartwood and leaf and named MuBL, MuHL and MuLL, respectively (Sá et al., 2009b; Napoleão et al., 2011, 2012). These lectins showed differences in carbohydrate content, specificity of carbohydrate binding site, activity at pH values and larvicidal activity. MuBL, MuHL and MuLL killed *A. aegypti* fourth-stage larvae (L4) with LC_{50} of 0.125, 0.04 and 0.202 mg/mL, respectively. The authors suggested that the larvicidal effect would be due to interaction of lectins with chitin and *N*-acetylglucosamine residues in glycosylated proteins at the surface of peritrophic matrix. Napoleão et al. (2012) also reported that MuLL retained its hemagglutinating activity after incubation with L4 digestive proteases for 96 h and attributed the resistance of MuLL to its glycoprotein nature.

5. EFFECTS OF PLANT TRYPSIN INHIBITORS ON LARVAL DEVELOPMENT

Proteinaceous trypsin inhibitors exert larvicidal effects on several species of economic importance. Larvae feeding on diets containing trypsin inhibitors showed decreased body weight, delayed growth and development, as well as low survival rate. The gene encoding the *Ipomoea batatas* trypsin inhibitor was introduced in tobacco plant to increase resistance to insect attack, and transgenic plants were used in larvicidal bioassays against *Spodoptera litura*. Growth of larvae was delayed and body weight was reduced. While leaves of the control tobacco plants were completely eaten after 4 days, transformed tobacco plants showed high degree of resistance to *S. litura* larvae, being observed only local lesions. The authors pointed out that transgenic expression of *I. batatas* trypsin inhibitor can be an efficient method of crop protection (Yeh et al., 1997). *S. litura* larvae were also sensitive to *Archidendron ellipticum* trypsin inhibitor (AeTI), when incorporated in larval diet. Early instar larva weights (first to third) were reduced by 43% and 61% after feeding on 100 µM and 150 µM AeTI, respectively. In addition, AeTI caused death (76%) of final instar larvae (Bhattacharyya et al., 2007).

Gomes et al. (2005a) showed that incorporation of trypsin inhibitor from *Cicer arietinum* seeds in diets resulted in mortality (43.3%) and delayed development for *Anthonomus grandis* larvae. Larvae of *Ephestia kuehniella* reared on diet containing the trypsin inhibitor from *Plathymenia foliolosa* seeds showed survival rate of 56%, larval mass reduction of 66%, increased metabolic cost, though efficiency of conversion of ingested food and digested food was reduced (Ramos et al., 2009). The trypsin inhibitor isolated from *Adenanthera pavonina* (ApTI) was also a larvicidal agent against *E. kuehniella* (Macedo et al., 2010). The development time of larvae was extended by 5 days and the pupal period was 10 days longer. In addition, the emergence of adults from larvae reared on ApTI diet was about only 28%, and survival of the adults that could emerge was also affected (38% mortality). ApTI also promoted mortality of *Callosobruchus maculatus* larvae (LD_{50} of 0.5%), and it was reported that for each 0.1% increase in the inhibitor, the mortality rate increases by 8% (Macedo et al., 2004b). *C. maculatus* larvae were also susceptible to *Crotalaria pallida* trypsin inhibitor, with values of LD_{50} (mortality) and ED_{50} (reduction in larvae mass) of 3.0% and 2.17%, respectively (Gomes et al., 2005b).

The Bowman-Birk proteinase inhibitor from red gram (*Cajanus cajan*) exerted deleterious effects against *Achaea janata* larvae. A dose-dependent decrease in body weight and survival rate with increasing concentration of inhibitors was observed. In all tested concentrations, the *C. cajan* inhibitor promoted 100% mortality of larvae after 20 days. While the larvae from control treatment weighed on average 171 mg after 6 days, mean weight of larvae fed on the inhibitor (at 8 μg/cm^2) was only 14 mg, corresponding to a 93% decrease. The *C. cajan* trypsin inhibitor was stable in presence of *A. janata* larval midgut proteinases (Prasad et al., 2010).

Protease inhibitors from the squash-type family have also showed insecticidal activities. The trypsin inhibitor from *Momordica charantia* seeds (McTI-II) was cloned and the recombinant McTI-II inhibited trypsin-like enzymes and total proteolytic activity from *Helicoverpa armigera* larvae by 81% and 70%, respectively. *H. armigera* larvae were fed on McTI-II incorporated into an artificial diet. On days 4, 8 and 12, the control larvae weighed respectively 67%, 82% and 74% more than McTI-II-fed larvae. Also, the ingestion of McTI-II resulted in 23% mortality in the larval population (Telang et al., 2009).

6. EFFECTS OF PLANT LECTINS AND TRYPSIN INHIBITORS ON LARVAL MORPHOLOGY

Morphologically, the midgut is the site most affected by plant lectins. The peritrophic matrix is a midgut structure that separates the contents of the gut lumen from the digestive epithelial cells, protecting against abrasion by plant cell wall fragments and infection by pathogens. In addition, it is involved in the compartmentalization of the digestive process. The peritrophic matrix is composed by chitin (polymer of *N*-acetylglucosamine) and glycoproteins such as peritrophins. The binding of lectins to these constituents leads to structural disorganization of peritrophic matrix with the consequent increase in permeability and damage to the microvillar brush border (Ryerse et al., 1994; Tellam et al., 1999; Hegedus et al., 2009).

The structural changes in the midgut of *Ostrinia nubilalis* larvae fed on wheat germ agglutinin (WGA) were investigated by Harper et al. (1998). In comparison with control, the larvae treated with WGA treatment showed smaller anterior mesenteron cells, lacked numerous vesicles, the food content in the gut lumen was scarce, and multilayered thick peritrophic matrices as well as numerous nascent matrices were present. Additionally, spherical and electro dense structures were observed and reported to be glycoproteins of the peritrophic matrix agglutinated by lectin. The authors suggested that matrix disruption allowed the dietary plant cell wall fragments to reach the brush border. Moreover, the secretion of new peritrophic matrices was an attempt to protect microvilli. However, this strategy did not prevent the penetration of bacilli as well as plant cell fragments in microvilli. The disorganization of peritrophic matrix resulted in disruption of brush border membrane and jeopardized insect's digestion and nutrition processes. WGA was also active against third-instar larvae of *Drosophila melanogaster*, promoting shortening, branching, swelling, distortion and disintegration of the midgut microvilli. These effects were similar to those observed in larvae kept starved, and the authors put forward the hypothesis that this lectin promoted a starvation-like effect on *D. melanogaster* larvae (Li et al., 2009).

Sauvion et al. (2004) reported structural changes induced by Con A, the lectin isolated from *Canavalia ensiformis* seeds, in the digestive tract of *Acyrthosiphon pisum* nymphs. Con A predominantly bound to luminal surface of stomach epithelial cells, promoting partial detachment, enlargement and deep digitation, with evidence of cell loss and abundant granular secretion into the lumen; larger amounts of cell debris and mucus were observed in the lumen, side by side with shedding of the striated border.

Analysis using inverted optical microscopy showed that *A. aegypti* larvae treated with WSMoL showed hypertrophy of segments and increased gut volume and disruption of the underlying epithelium that delimits the gut. Thus, the authors suggested that the absence of underlying epithelium may indicate that larvicidal activity of WSMoL was probably due to damage to the larvae midgut (Coelho et al., 2009).

Trypsin inhibitors also promoted morphological changes in insects. Damaged morphology of larvae as well as of pupae and adults was described for *Anthonomus grandis* larvae fed on diet containing trypsin inhibitors of soybean or *Cicer arietinum* (Franco et al., 2003; Gomes et al., 2005a). The deformities included absence of wings and thorax, and it was suggested that this effect was due to the inhibition of proteases involved in the metamorphosis process, mainly in the production of new adult tissues and enzyme systems.

7. MODULATION OF ENZYME ACTIVITIES FROM LARVAL GUT BY PLANT LECTINS AND TRYPSIN INHIBITORS

Lectins are able to block the activity of enzymes by binding to the sugar moiety in glycosylated enzymes, or by binding to sites other than the substrate binding site in non-glycosylated enzymes (Macedo et al., 2007). Larvae of *Lacanobia oleracea* fed on semi-artificial diet supplemented with *Galanthus nivalis* agglutinin (GNA) and Con A showed increased levels of aminopeptidase and trypsin activities as well as slightly increased alkaline phosphatase activity. GNA also induced an increase in α-glucosidase activity. However, these induced stimulations were not maintained in the long term (Fitches and Gatehouse, 1998).

The larvicidal lectin from *Myracrodruon urundeuva* leaf (MuLL) inhibited the activity of proteases and trypsin-like enzymes from *Aedes aegypti* L4 gut, and zymography revealed that the intensity of the lytic zones for polypeptides of approximately 14.2 and 16.9 kDa were reduced (Napoleão et al., 2012). The trypsin-like activity from the midgut of *Ephestia kuehniella* larvae also decreased after consumption of an artificial diet containing *Annona coriacea* lectin (ACLEC); on the other hand, it was observed a high increase (78%) in the tryptic activity in faecal extracts. The authors suggested that the presence of protease activity in the feces may be due to perturbation of the organization of the peritrophic membrane leading to enzyme decompartmentalization (Coelho et al., 2007).

Treatment of *Callosobruchus maculatus* larvae with *Bauhinia monandra* leaf lectin (BmoLL) did not alter the papain-like activity in the midgut, though it led to a reduction in α-amylase activity. However, incubation of BmoLL with larvae midgut homogenates increased α-amylase activity (Macedo et al., 2007). Similarly, *M. urundeuva* leaf lectin also exerted a stimulatory effect on α-amylase activity in gut of *A. aegypti* L4 (Napoleão et al., 2012).

Midgut of insects of several orders contains trypsin-like enzymes, and these proteins are potential targets for insecticidal trypsin inhibitors. The blocking of the activity of serine proteases in larva gut leads to impairment of digestion and absorption of essential amino acids. The trypsin inhibitor from *Cicer arietinum* seeds and the soybean Kunitz trypsin inhibitor inhibited the activity of serine proteinases from the midgut of *Anthonomus grandis* larvae (Franco et al., 2003; Franco et al., 2004). The *Cajanus cajan* protease inhibitor showed remarkable inhibitory activity against midgut trypsin-like enzymes of *Achaea janata* (Prasad et al., 2010), and enzymes from *Callosobruchus maculatus* gut were strongly susceptible to *Crotalaria pallida* trypsin inhibitor (CpaTI) in an *in vitro* assay, inhibiting enzyme activity by 74.4% (Gomes et al., 2005b). The insect organism can react by increasing enzyme synthesis in order to compensate for the drop in enzymatic activity due to the action of the inhibitor. The feeding on diets containing *Adenanthera pavonina* trypsin inhibitor (ApTI) resulted in massive overproduction of cysteine proteinases by *Callosobruchus maculatus* larvae. However, the increased synthesis of these enzymes led to a shortage of essential amino acids required for the production of other proteins (Macedo et al., 2004b). Kinetic analyses of the inhibition of *Spodoptera litura* midgut trypsin-like activity by *Archidendron elipticum* trypsin inhibitor (AeTI) demonstrated a competitive mechanism and yielded a K_i of 1.5 nM, indicating a high affinity of AeTI for the insect trypsin-like enzymes. Interestingly, the trypsin-like activity in the midgut of *S. litura* larvae fed on 150 µM AeTI decreased significantly in comparison with larvae fed on control diet instead of an expected high protease production (Bhattacharyya et al., 2007). Insects are also able to modify their set of digestive proteinases in response to anti-metabolic effects of inhibitors; in this sense, in order to be a good insecticide candidate, a trypsin inhibitor should not act very specifically against one or another class of enzyme that can be expressed, which would prevent the selection of a resistant protease form. *C. maculatus* larvae were not able to adapt to the ingestion of ApTI, since no new protease forms resistant to ApTI were induced, at least in first-, second-, third and fourth-instar larvae of the first generation; similarly, no new proteolytic form was induced in the gut of *E. kuehniella* larvae after rearing on diet containing 1% ApTI (Macedo et al., 2004b; Macedo et al., 2010). In addition, these inhibitors can be applied in combination with another inhibitor or insecticide. For example, protease inhibitors have been shown to synergize the effect of *Bacillus thuringiensis* toxin (Jaber et al., 2010).

CONCLUSION

This chapter reports plants as sources of environmentally-friendly proteins with deleterious effects against immature forms of insects. Lectins affect larval survival, weight, feeding and duration of development, and promote damage to midgut structures such as peritrophic matrix and brush border membrane. Proteinaceous trypsin inhibitors promote larval mortality, deregulation of protease activity in the larva midgut and morphological abnormalities in pupae and adults derived from larvae fed on diets supplemented with trypsin inhibitor. These negative effects on insect biology can prevent the emergence of adults, and thus lectins and trypsin inhibitors have potential use in strategies for control of pests and disease vectors.

ACKNOWLEDGMENTS

The authors express their gratitude to the *Conselho Nacional de Desenvolvimento Científico e Tecnológico* (CNPq) for research grants and fellowships (P.M.G. Paiva and L.C.B.B. Coelho). We are also grateful to the *Fundação de Amparo à Ciência e Tecnologia do Estado de Pernambuco* (FACEPE) and the *Coordenação de Aperfeiçoamento de Pessoal de Nível Superior* (CAPES) for financial support. E.V. Pontual and T.H. Napoleão would like to thank FACEPE and CAPES, respectively, for graduate scholarships. We thank Felix Nonnenmacher for English editing and Maria Barbosa Reis da Silva for technical assistance.

REFERENCES

Allotey, J. and Azalekor, W. (2000). Some aspects of the biology and control using botanicals of the rice moth, *Corcyra cephalonica* (Stainton), on some pulses. *Journal of Stored Products Research,* 36, 235-243.

Ambrosi, M., Cameron, N.R., and Davis. B.G. (2005). Lectins: tools for the molecular understanding of the glycocode. *Organic and Biomolecular Chemistry,* 3, 1593-1608.

Azab, S.G., Sadek, M.M., and Crailsheim, K. (2001). Protein metabolism in larvae of the cotton leaf-worm *Spodoptera littoralis* (Lepidoptera: Noctuidae) and its response to three mycotoxins. *Environmental Entomology,* 30, 817-823.

Bandyopadhyay, S., Roy, A., and Das, S. (2001). Binding of garlic (*Allium sativum*) leaf lectin to the gut receptors of homopteran pests is correlated to its insecticidal activity. *Plant Science,* 161, 1025-1033.

Bhat, G.G., Shetty, K.N., Nagre, N.N., Neekhra, V.V., Lingaraju, S., Bhat, R.S., Inamdar, S.R., Suguna, K., and Swamy, B.M. (2010). Purification, characterization and molecular cloning of a monocot mannose-binding lectin from *Remusatia vivipara* with nematicidal activity. *Glyconjugate Journal,* 27, 309-320.

Bhattacharyya, A., Leighton, S.M., and Babu, C.R. (2007). Bioinsecticidal activity of *Archidendron ellipticum* trypsin inhibitor on growth and serine digestive enzymes during larval development of *Spodoptera litura. Comparative Biochemistry and Physiology,* Part C, 145, 669-677.

Blackford, M. and Dinan, L. (1997). The tomato moth *Lacanobia oleracea* (Lepidoptera, Noctuidae) detoxifies ingested 20-hydroxyecdysone, but is susceptible to the ecdysteroid agonists RH-5849 and RH-5992. *Insect Biochemistry and Molecular Biology,* 27, 167-177.

Bode, W. and Huber, R. (2000). Structural basis of the endoproteinase–protein inhibitor interaction. *Biochimica et Biophysica Acta - Protein Structure and Molecular Enzymology,* 1477, 241-252.

Bray, E.A., Bailey-Serres, J., and Weretilnyk, E. (2000). Responses to abiotic stresses, In: Buchanan, B.B.; Gruissen, W., and Jones L.R. (Eds.) Biochemistry and Molecular Biology of Plants, *American Society of Plant Physiologists,* New York, pp. 1158-1203.

Chen, P., Liu, Y., Kang, X., Sun, L., Yang, P., and Tang, Z. (2008), Identification of N - glycan of alpha-fetoprotein by lectin affinity microarray. *Journal of Cancer Research and Clinical Oncology,* 134, 851-860.

Coats, J.R. (1990). Mechanisms of toxic actions and structure-activity relationships for organochlorine and synthetic pyrethroid insecticides. *Environmental Health Perspectives,* 87, 255-262.

Coelho, J.S., Santos, N.D.L., Napoleão, T.H., Gomes, F.S., Ferreira, R.S., Zingali, R.B., Coelho, L.C.B.B., Leite, S.P., Navarro, D.M.A.F., and Paiva, P.M.G. (2009). Effect of *Moringa oleifera* lectin on development and mortality of *Aedes aegypti* larvae. *Chemosphere,* 77, 934-938.

Coelho, M.B., Marangoni, S., and Macedo, M.L.R. (2007). Insecticidal action of *Annona coriacea* lectin against the flour moth *Anagasta kuehniella* and the rice moth *Corcyra cephalonica* (Lepidoptera: Pyralidae). *Comparative Biochemistry and Physiology,* Part C, 146, 406-414.

Correia, M.T.S, Coelho, L.C.B.B., and Paiva, P.M.G. (2008). Lectins, carbohydrate recognition molecules: Are they toxic? In: Siddique, Y.H. (Ed.) Recent Trends in Toxicology, vol. 37, *Transworld Research Network,* Kerala, pp. 47-59.

Fang, E.F., Lin, P., Wong, J.H., Tsao, S.W., and Ng, T.B. (2010). A lectin with anti-HIV-1 reverse transcriptase, antitumor, and nitric oxide inducing activities from seeds of *Phaseolus vulgaris* cv. extralong autumn purple bean. *Journal of Agricultural and Food Chemistry,* 58, 2221-2229.

Fernandes, M.P., Inada, N.M., Chiaratti, M.R., Araújo, F.F.B., Meirelles, F.V., Correia, M.T.S., Coelho, L.C.B.B., Alves, M.J.M., Gadelha, F.R., and Vercesi, A.E. (2010). Mechanism of *Trypanosoma cruzi* death induced by *Cratylia mollis* seed lectin. *Journal of Bioenergetics and Biomembranes,* 42, 69-78.

Ferreira, R.S., Napoleão, T.H., Santos, A.F.S., Sá, R.A., Carneiro-da-Cunha, M.G., Morais, M.M.C., Silva-Lucca, R.A., Oliva, M.L.V., Coelho, L.C.B.B., and Paiva, P.M.G. (2011) Coagulant and antibacterial activities of the water-soluble seed lectin from *Moringa oleifera*. *Letters in Applied Microbiology,* 53, 186-192.

Fitches, E., Gatehouse, A.M.R., and Gatehouse, J.A. (1997). Effects of snowdrop lectin (GNA) delivered via artificial diet and transgenic plants on the development of tomato moth (*Lacanobia oleracea*) larvae in laboratory and glasshouse trials. *Journal of Insect Physiology*, 43, 727-739.

Fitches, E., and Gatehouse, J.A. (1998). A comparison of the short and long term effects of insecticidal lectins on the activities of soluble and brush border enzymes of tomato moth larvae (*Lacanobia oleracea*). *Journal of Insect Physiology,* 44, 1213-1224.

Fitches, E., Woodhouse, S.D., Edwards, J.P., and Gatehouse, J.A. (2001). In vitro and in vivo binding of snowdrop (*Galanthus nivalis* agglutinin; GNA) and jackbean (*Canavalia ensiformis*; Con A) lectins within tomato moth (*Lacanobia oleracea*) larvae; mechanisms of insecticidal action. *Journal of Insect Physiology,* 47, 777-787.

Forsberg, A. and Puu, G. (1984). Kinetics for the inhibition of acetylcholinesterase from the electric eel by some organophosphates and carbamates. *European Journal of Biochemistry,* 140, 153-156.

Franco, O., Santos, R.C., Batista, J.A.N., Mendes, A.C.M., Araújo, M.A.M., Monnerat, R.G., Grossi-de-Sá, M.F., and Freitas, S.M. (2003). Effects of black-eyed pea trypsin/chymotrypsin inhibitor on proteolytic activity and on development of *Anthonomus grandis*. *Phytochemistry,* 63, 343-349.

Franco, O.L., Dias, S.C., Magalhães, C.P., Monteiro, A.C.S., Bloch Jr, C., Melo, F.R., Oliveira-Neto, M.R., Monnerat, R.G., and Grossi-de-Sá, M.G. (2004). Effects of soybean

Kunitz trypsin inhibitor on the cotton boll weevil (*Anthonomus grandis*). *Phytochemistry*, 65, 81-89.

Gemeiner, P., Mislovičová, D., Tkáč, J., Švitel, J., Pätoprstý, V., Hrabárová, E., Kogan, G., and Kožár, T. (2009). Lectinomics II. A highway to biomedical/clinical diagnostics. *Biotechnology Advances* 27, 1-15.

Gibbons, S. (2004) Anti-staphylococcal plant natural products. *Natural Product Reports*, 21, 263–277.

Gomes, A.P.G., Dias, S.C., Bloch Jr, C., Melo, F.R., Furtado Jr, J.R., Monnerat, R.G., Grossi-de-Sá, M.G., and Franco, O.L. (2005a). Toxicity to cotton boll weevil *Anthonomus grandis* of a trypsin inhibitor from chickpea seeds. *Comparative Biochemistry and Physiology*. Part B, 140, 313-319.

Gomes, C.E.M., Barbosa, A.E.A.D., Macedo, L.L.P., Pitanga, J.C.M., Moura, F.T., Oliveira, A.S., Moura, R.M., Queiroz, A.F.S., Macedo, F.P., Andrade, L.B.S., Vidal, M.S., and Sales, M.P. (2005b). Effect of trypsin inhibitor from *Crotalaria pallida* seeds on *Callosobruchus maculatus* (cowpea weevil) and *Ceratitis capitata* (fruit fly). *Plant Physiology and Biochemistry*, 43, 1095-1102.

Hamshou, M., Smagghe, G., Shahidi-Noghabi, S., De Geyter, E., Lannoo, N., and Van Damme E.J.M. (2010). Insecticidal properties of *Sclerotinia sclerotiorum* agglutinin and its interaction with insect tissues and cells. *Insect Biochemistry and Molecular Biology*, 40, 883-890.

Harper, M.S., Hopkins, T.L., and Czapla, T.H. (1998). Effect of wheat germ agglutinin on formation and structure of the peritrophic membrane in European corn borer (*Ostrinia nubilalis*) larvae. *Tissue and Cell*, 30, 166-176.

Hegedus, D., Erlandson, M., Gillott, C., and Toprak, U. (2009). New Insights into Peritrophic Matrix Synthesis, Architecture, and Function. *Annual Review of Entomology*, 54, 285-302.

Hemingway, J. and Ranson, H. (2000). Insecticide resistance in insect vectors of human disease. *Annual Review of Entomology*, 45, 371-391.

Jaber, K., Haubruge, É., and Francis, F. (2010). Development of entomotoxic molecules as control agents: illustration of some protein potential uses and limits of lectins (Review). *Biotechnology, Agronomy, Society and Environment*, 14, 225-241.

Jallow, M.F.A., Cunningham, J.P., and Zalucki, M.P. (2004). Intra-specific variation for host plant use in *Helicoverpa armigera* (Hubner) (Lepidoptera: Noctuidae): implications for management. Crop Protection, 23, 955-964.

Kennedy, J.F., Paiva, P.M.G., Correia, M.T.S., Cavalcanti, M.S.M., and Coelho, L.C.B.B. (1995). Lectins, versatile proteins of recognition: a review. *Carbohydrate Polymers*, 26, 219-230.

Laskowski Jr., M. and Qasim, M. (2000). What can the structures of enzyme-inhibitor complexes tell us about the structures of enzyme substrate complexes? *Biochimica et Biophysica Acta*, 1477, 324-337.

Li, H.M., Sun, L., Mittapalli, O., Muir, W.M., Xie, J., Wu, J., Schemerhorn, B.J., Sun, W., Pittendrigh, B.R., and Murdock, L.L. (2009). Transcriptional signatures in response to wheat germ agglutinin and starvation in *Drosophila melanogaster* larval midgut. *Insect Molecular Biology*, 18, 21–31.

Liao, H., Ren, W., Kang, Z., Zhao, X.J., and Du, L.F. (2007). A trypsin inhibitor from *Cassia obtusifolia* seeds: isolation, characterization and activity against *Pieris rapae*. *Biotechnology Letters*, 29, 653-658.

Macedo, M.L.R., Damico, D.C.S., Freire, M.G.M., Toyama, M.H., Marangoni, S., and Novello, J.C. (2003). Purification and characterization of an N-Acetylglucosamine-binding lectin from *Koelreuteria paniculata* seeds and its effect on the larval development of *Callosobruchus maculatus* (Coleoptera: Bruchidae) and *Anagasta kuehniella* (Lepidoptera: Pyralidae). *Journal of Agricultural and Food Chemistry*, 51, 2980-2986.

Macedo, M.L.R., Castro, M.M., and Freire, M.G.M. (2004a). Mechanisms of the insecticidal action of TEL (*Talisia esculenta* lectin) against *Callosobruchus maculatus* (Coleoptera: Bruchidae). *Archives of Insect Biochemistry and Physiology*, 56, 84–96.

Macedo, M.L.R., Sá, C.M., Freire, M.G.M., and Parra, J.L.P. (2004b). A Kunitz-type inhibitor of coleopteran proteases, isolated from *Adenanthera pavonina* L. seeds and its effect on *Callosobruchus maculatus*. *Journal of Agricultural and Food Chemistry*, 52, 2533-2540.

Macedo, M.L.R., Freire, M.G.M., Silva, M.B.R., and Coelho, L.C.B.B. (2007). Insecticidal action of *Bauhinia monandra* leaf lectin (BmoLL) against *Anagasta kuehniella* (Lepidoptera: Pyralidae), Zabrotes subfasciatus and *Callosobruchus maculatus* (Coleoptera: Bruchidae). *Comparative Biochemistry and Physiology*, Part A, 146, 486-498.

Macedo, M.L.R., Durigan, R.A., Silva, D.S., Marangoni, S., Freire, M.G.M., and Parra, J.R.P. (2010). *Adenanthera pavonina* trypsin inhibitor retard growth of *Anagasta kuehniella* (Lepidoptera: Pyralidae). *Archives of Insect Biochemistry and Physiology*, 73, 213-231.

Major, I.T. and Constabel, P. (2008). Functional analysis of the kunitz trypsin inhibitor family in poplar reveals biochemical diversity and multiplicity in defense against herbivores multiplicity in defense against herbivores. *Plant Physiology*, 146, 888-903.

McKinlay, R., Plant, J.A., Bell, J.N.B., and Voulvoulis, N. (2008). Endocrine disrupting pesticides: Implications for risk assessment. *Environment International*, 34, 168-183.

Melo C.M.L., Porto, C.S., Melo-Júnior, M.R., Mendes, C. M., Cavalcanti, C.C. B., Coelho, L.C.B.B., Porto, A.L.F., Leão, A.M.A.C., and Correia, M.T.S. (2011). Healing activity induced by Cramoll 1,4 lectin in healthy and immunocompromised mice. *International Journal of Pharmaceutics*, 408, 113-119.

Michiels, K., Van Damme, E.J., and Smagghe, G. (2010). Plant-insect interactions: what can we learn from plant lectins? *Archives of Insect Biochemistry and Physiology*, 73, 193-212.

Migliolo, L., Oliveira, A.S., Santos, E.A., Franco, O.L., and Sales, M.P. (2010). Structural and mechanistic insights into a novel non-competitive Kunitz trypsin inhibitor from *Adenanthera pavonina* L. seeds with double activity toward serine and cysteine-proteinases. *Journal of Molecular Graphics and Modelling*, 29, 148-156.

Murdock, L.L., Huesing, J.E., Nielsen, S.S., Pratt, R.C., and Shade, R.E. (1990). Biological effects of plant lectins on the cowpea weevil. *Phytochemistry*, 29, 85-89.

Napoleão, T.H., Gomes, F.S., Lima, T.A., Santos, N.D.L., Sá, R.A., Albuquerque, A.C., Coelho, L.C.B.B., and Paiva, P.M.G. (2011). Termiticidal activity of lectins from *Myracrodruon urundeuva* against *Nasutitermes corniger* and its mechanisms. *International Biodeterioration and Biodegradation*, 65, 52-59.

Napoleão, T.H., Pontual, E.V., Lima, T.A., Santos, N.D.L., Sá, R.A., Coelho, L.C.B.B., Navarro, D.M.A.F., Paiva, P.M.G. (2012). Effect of *Myracrodruon urundeuva* leaf lectin on survival and digestive enzymes of *Aedes aegypti* larvae. *Parasitology Research,* doi:10.1007/s00436-011-2529-7.

Nauen, R. (2007). Perspective insecticide resistance in disease vectors of public health importance. *Pest Management Science,* 63, 628-633.

Oliveira, C.F.R., Luz, L.A., Paiva, P.M.G., Coelho, L.C.B.B., Marangoni, S., and Macedo, M.L.R. (2011). Evaluation of seed coagulant *Moringa oleifera* lectin (cMoL) as a bioinsecticidal tool with potential for the control of insects. *Process Biochemistry,* 46, 498-504.

Oliveira, M.D.L., Andrade, C.A.S., Magalhães, N.S., Coelho, L.C.B.B., Teixeira, J.A., Carneiro-da-Cunha, M.G., and Correia, M.T.S. (2008). Purification of a lectin from *Eugenia uniflora* L. seeds and its potential antibacterial activity. *Letters in Applied Microbiology,* 46, 371-376.

Peumans, W.J. and Van Damme, E.J.M. (1998). Plant lectins: versatile proteins with important perspectives in biotechnology. *Biotechnology and Genetic Engineering Reviews,* 15, 199-227.

Prasad, E.R., Dutta-Gupta, A., and Padmasree, K. (2010). Insecticidal potential of Bowman–Birk proteinase inhibitors from red gram (*Cajanus cajan*) and black gram (*Vigna mungo*) against lepidopteran insect pests. *Pesticide Biochemistry and Physiology,* 98, 80-88.

Ramos VS, Freire MGM, Parra JRP, and Macedo MLR. (2009). Regulatory effects of an inhibitor from Plathymenia foliolosa seeds on the larval development of *Anagasta kuehniella* (Lepidoptera). *Comparative Biochemistry and Physiology* A, 152, 255-261.

Rodriguez-Navarro, D.N., Dardanelli, M.S., and Ruíz-Saínz, J.E. (2007). Attachment of bacteria to the roots of higher plants. *FEMS Microbiology Letters,* 272, 127-136.

Roy, S., Mandal, C., and Dutta, S.K. (2011). Site-directed mutagenesis to identify key residues in structure-function relationship of winged bean chymotrypsin-trypsin inhibitor and 3-D structure prediction. *Protein and Peptide Letters,* 18, 471-479.

Rüdiger, H. and Gabius, H.J. (2001). Plant lectins: occurrence, biochemistry, functions and applications. *Glycoconjugate Journal,* 18, 589-613.

Ryerse, J.S., Purcell, J.P., and Sammons, R.D. (1994). Structure and formation of the peritrophic membrane in the larva of the southern corn rootworm, *Diabrotica undecimpunctata. Tissue and Cell,* 26, 431-437.

Sá, R.A., Napoleão, T.H., Santos, N.D.L., Gomes, F.S., Albuquerque, A.C., Xavier, H.S., Coelho, L.C.B.B., Bieber, L.W., and Paiva, P.M.G. (2008). Induction of mortality on *Nasutitermes corniger* (Isoptera, Termitidae) by *Myracrodruon urundeuva* heartwood lectin. *International Biodeterioration and Biodegradation,* 62, 460-464.

Sá, R.A., Gomes, F.S., Napoleão, T.H., Santos, N.D.L., Melo, C.M.L., Gusmão, N.B., Coelho, L.C.B.B., Paiva, P.M.G., and Bieber, L.W. (2009a). Antibacterial and antifungal activities of *Myracrodruon urundeuva* heartwood. *Wood Science and Technology,* 43, 85-95.

Sá, R.A., Santos, N.D.L., Silva, C.S.B., Napoleão, T.H., Gomes, F.S., Cavada, B.S., Coelho, L.C.B.B., Navarro, D.M.A.F., Bieber, L.W., and Paiva, P.M.G. (2009b). Larvicidal activity of lectins from *Myracrodruon urundeuva* on *Aedes aegypti. Comparative Biochemistry and Physiology,* Part C, 149, 300-306.

Sadeghi, A., Smagghe, G., Broeders, S., Hernalsteens, J., De Greve, H., Peumans, W.J., and Van Damme, E.J.M. Ectopically expressed leaf and bulb lectins from garlic (*Allium sativum* L.) protect transgenic tobacco plants against cotton leafworm (*Spodoptera littoralis*). *Transgenic Research,* 17- 9-18.

Santos, A.F.S., Luz, L.A., Argolo, A.C.C., Teixeira, J.A., Paiva, P.M.G., and Coelho, L.C.B.B. (2009). Isolation of a seed coagulant *Moringa oleifera* lectin. *Process Biochemistry,* 44, 504–508.

Sauvion, N., Nardonb, G., Febvay, G., Gatehouse, A.M.R., and Rahbé, Y. (2004). Binding of the insecticidal lectin Concanavalin A in pea aphid *Acyrthosiphon pisum* (Harris) and induced effects on the structure of midgut epithelial cells. *Journal of Insect Physiology,* 50, 1137–1150.

Souza, J.D., Silva, M.B.R., Argolo, A.C.C., Napoleão, T.H., Sá, R.A., Correia, M.T.S., Paiva, P.M.G., Silva, M.D.C., and Coelho, L.C.B.B. (2011). A new *Bauhinia monandra* galactose-specific lectin purified in milligram quantities from secondary roots with antifungal and termiticidal activities. *International Biodeterioration and Biodegradation,* 65, 696-702.

Telang, M.A., Pyati, P., Sainani, M., Gupta, V.S., and Giri, A.P. (2009). *Momordica charantia* trypsin inhibitor II inhibits growth and development of *Helicoverpa armigera*. *Insect Science,* 16, 371-379.

Tellam, R.L., Wijffels, G., and Wiladsen, P. (1999). Peritrophic matrix proteins. *Insect Biochemistry and Molecular Biology,* 29, 87-101.

World Health Organization (2008). *Chikungunya.* Fact sheet 327.

World Health Organization (2009a) *Yellow Fever.* Fact sheet 100.

World Health Organization (2009b). *Dengue and dengue haemorrhagic fever.* Fact sheet 117.

Yan, Q., Zhu, L., Kumar, N., Jiang, Z., Huang, L. (2010). Characterisation of a novel monomeric lectin (AML) from *Astragalus membranaceus* with anti-proliferative activity. *Food Chemistry,* 122, 589-595.

Yeh, K.-W., Lin, M.-I., Tuan, S.-J., Chen, Y.-M., Lin, C.-J., and Kao, S.-S. (1997). Sweet potato (*Ipomoea batatas*) trypsin inhibitors expressed in transgenic tobacco plants confer resistance against Spodoptera litura. *Plant Cell Reports,* 16, 696-699.

Zhu-Salzman, K., Shade, R.E., Koiwa, H., Salzman, R.A., Narasimhan, M., Bressan, R.A., Hasegawa, P.M., and Murdock, L.L. (1998). Carbohydrate binding and resistance to proteolysis control insecticidal activity of *Griffonia simplicifolia* lectin II. *Proceedings of the National Academy of Sciences of the United States,* 95, 15123–15128.

In: Larvae: Morphology, Biology and Life Cycle
Editors: Kia Pourali and Vafa Niroomand Raad

ISBN: 978-1-61942-662-7
© 2012 Nova Science Publishers, Inc.

Chapter 4

INTRANNUAL EFFECTS OF BIOTIC AND ABIOTIC FACTORS ON GROWTH AND MORTALITY OF ANADROMOUS TWAITE SHAD, *ALOSA FALLAX FALLAX* (LACÉPÈDE, 1803), LARVAE

E. Esteves and J. P. Andrade*

Centro de Ciências do Mar do Algarve CCMar – CIMAR Laboratório Associado,
Universidade do Algarve, Campus de Gambelas, Faro, Portugal

ABSTRACT

The anadromous twaite shad, *Alosa fallax fallax* (Lacépède, 1803), is a European Union Berne Convention *Natura 2000* species, requiring protection from range states. In Portugal it is classified as "Vulnerable" by the Instituto de Conservação da Natureza using IUCN criteria but still migrates upriver to spawn in several rivers, *e.g.* River Mira. Evidence suggests that this species exhibits homing behavior. This stock-river relationship is important considering that environmental conditions during the embryo-larval period in the freshwater reaches of rivers play an important role in the future of populations, namely through their impact on growth and mortality rates that affect stage duration and survival. Nonetheless, published information on growth and/or mortality of twaite shad larvae is anecdotal. This study was designed to model growth and mortality rates and study their within season variation and examine to what extent short-term changes in environmental and biological factors affect within-year changes in growth rates and relative survival of larvae of twaite shad in the River Mira.

Plankton samples were collected every other week from 28 March until 7 July 2000, the presumed spawning season, at stations in the upper River Mira. Shad larvae were sorted out of the zooplankton. Microstructure of *sagittae* from *c.* 330 post yolk-sac shad larvae, 7.21 to 20.40 mm standard length, was analysed using an image analysis system and an age-length key was prepared. Growth in length of larvae, g^*, was estimated based on the backcalculated length-at-age using a model-based resampling (bootstrap) approach. Instantaneous daily mortality rates, Z, of larvae were estimated from the ln-

---

* Present address: Eduardo Esteves, Instituto Superior de Engenharia, Universidade do Algarve, Campus da Penha, 8005-139 Faro, Portugal. E-mail: eesteves@ualg.pt.

transformed form of an exponential model of decline in abundance of adjusted numbers with respect to age (time). The relationship among nine environmental and biological variables, and instantaneous growth and mortality rates of shad larvae was assessed.

Larvae do not grow at the same rate or experience uniform mortality throughout the spawning season. They were estimated to grow at a faster rate earlier in the season when compared to individuals collected later, after mid-May. Early and later-spawned larvae were subject to comparable and declining mortality rates. Seasonal changes in river (water) temperature and rainfall (a proxy of river discharge) as well as variation in crustacean *nauplii* abundance and feeding incidence, coincident with early larval development, are important variables, regulating the growth and mortality of larval twaite shad.

INTRODUCTION

The anadromous twaite shad, *Alosa fallax fallax* (Lacépède, 1803), is a *Natura 2000* species listed in the European Union Berne Convention thus requiring protection from member states. In Portugal it is classified as "Vulnerable" by the Instituto de Conservação da Natureza using IUCN criteria [1] but still migrates upriver to spawn in several rivers along Portugal, namely in the Rivers Minho, Mondego, Tejo, Mira and Guadiana. Evidence suggests that this species exhibits homing behaviour [2, 3]. This stock-river relationship is important considering that environmental conditions during the embryo-larval period in the freshwater reaches of rivers play an important role in the future of populations [4], namely through their impact on growth and mortality rates that affect stage duration and survival.

Small changes in growth and survival rates during early life-history of fish are believed to have significant effects on later recruitment success [5]. In fact, the central theme of Hjort's critical period hypothesis and Cushing's match-mismatch hypothesis is that food availability and/or larval predation are the proximate determinants of larval survival and the later year-class strength, whereas hydrographical events are the ultimate causes that drive the system [6]. Nonetheless, published information on growth and/or mortality of twaite shad larvae is anecdotal [see 7 for recent review], consisting mainly of sizes-at-hatch or sizes-at-age and qualitative assessments [8, 9, 10]. In contrast, those traits have been investigated on the early life-history stages of closely-related species, e.g. allis shad *A. alosa* (L.) in the Gironde-Garonne-Dordogne watershed (France) [11, 12] and the American shad *A. sapidissima* (Wilson, 1811) in the Connecticut River (USA) [13, 14, 15] and in the Hudson River (USA) [16, 17]. Leach and Houde [18] studied experimentally the effects of environmental factors on growth survival and production of American shad larvae. Letcher et al. [19] carried out laboratory experiments to study the differences in consumption and growth rates of Lake Michigan (USA) larvae including alewife *A. pseudoharengus* (Wilson, 1811).

This chapter reports on a study designed to: 1) model growth and mortality rates and study their within-season variation; and 2) examine to what extent short-term changes in environmental and biological factors affect within-year changes in relative growth rate and survival of larvae of twaite shad *A. f. fallax* in the River Mira (SW Portugal).

STUDY AREA

The River Mira is a 145-km long, relatively narrow (maximum width <400 m near the mouth and does not exceed 30 m upriver) and shallow (maximum depth <10 m near the mouth but varies between 0.5 and 4 m at the upper limit of tidal influence) watercourse located in SW Portugal (Figure 1). Tidal influence extends to approximately 40 km from the mouth (Odemira), where tide reversal and salinity changes are observed. Therein, the river bed is composed of mixed sediments (very fine silt up to coarse sand, pebbles and cobbles), submerged vegetation and abundant debris. The distribution of rainfall is markedly seasonal, with c. 80% of the annual precipitation (mean of 666 mm) concentrated in the period October-March [20]. Since 1968, the Santa Clara Dam has been used for local water supply and irrigation, and energy production, and largely contributed to the regulation of river discharge despite the skewed precipitation. Nonetheless, average monthly discharge increases from 1.3×10^9 m^3 in October to 26.7×10^9 m^3 during December and then decreases to 2.5×10^9 m^3 in March/April (no discharge has been registered from June to September [20]).

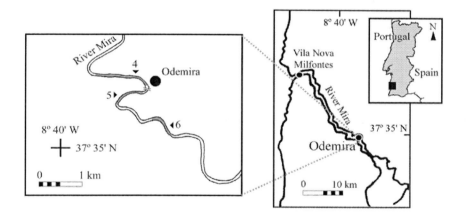

Figure 1. Location of sampling stations no. 4 – 6 in the upper River Mira (SW Portugal).

DATA COLLECTION AND SAMPLE PROCESSING

Plankton samples were collected every other week from 28 March (day 88 after 1 January) until 7 July 2000 (day 189) at three stations (no. 4 – 6) in the upper River Mira (Figure 1) with a conical net (0.37 m diameter, 1.6 m long, 0.5 mm mesh size) towed 1 m below the surface, at a constant speed of approximately 1 m s^{-1} for 10 minutes. Sampling took place during daylight hours (9 a.m. to 7 p.m.) and the tows were performed in a zigzag path against dominant flow trying to encompass evenly the margins and the main channel of the river. A flowmeter (Hydro-Bios Apparatebau GmbH, Germany) was attached inside the net to estimate the water volume filtered (median = 62.8 m^3, range = 14.9–113.1 m^3). The water surface temperature (°C) was measured using a hand-held thermometer.

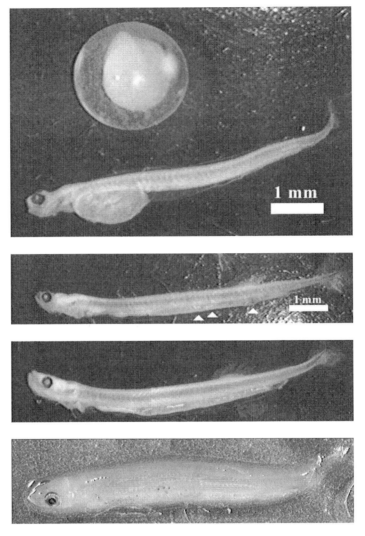

Figure 2. Images of (from top to bottom) 2.17 mm-diameter twaite shad egg, 6.42 mm L_S yolk-sac larvae, 8.23 mm L_S larvae with gut content (triangles), 10.11 mm L_S larvae and 19.90 mm L_S pre-metamorphosis larvae obtained with a CCD camera connected to a stereoscope.

Daily rainfall (mm), a proxy of river flow/discharge, was obtained from the Sistema Nacional de Informação de Recursos Hídricos [20] and standard errors of moving averages (prior 7-days) were used as an estimate of rainfall variability instead of actual, daily values because precipitation in the upper River Mira area varies erratically.

Zooplankton samples were sorted in a black tray, the shad larvae (Figure 2) retrieved and immediately frozen in liquid nitrogen (-197 °C) for complementary biochemical analysis [21]. Zooplankters were stored in buffered 4% formaldehyde solutions and later were identified using published information [22, 23, 24, 25, 26]. In this chapter, taxa were grouped as "crustacean *nauplii*", "zooplankton" (*viz.* isopods, small insects, cladocerans and copepodites) and "predators" (taxa that act as predators of shad larvae and/or competitors for same ecological niche, *i.e.* small medusae, mysids and amphipods). Plankton density was expressed as number of individuals per 100 m^3 of water (n 100 m^{-3}). Moreover, the Shannon index of diversity H' [27] was calculated for each sample.

Microplankton was sampled using a 63 μm mesh size sieve towed vertically from 1 m depth and its biomass was expressed as ash-free dry weight (AFDW) per cubic meter of water (mg AFDW m^{-3}).

During shad larvae processing for otolith analysis (see below) the presence/absence of visible items in the gut of twaite shad larvae was noted (Figure 2) and used to estimate feeding incidence, *i.e.* the proportion of shad larvae with gut content $p_W = n_W/n$ where n_W is the number of shad with visible gut content and n is the number of larvae in a sample.

OTOLITH ANALYSIS, AGEING AND HATCH DATE DISTRIBUTION

Otoliths were removed from twaite shad larvae using a combination of lightly concentrated 4.5% NaOH rinse (<3 min) and fine-needle manipulation under a stereoscope (Figure 3). They were then mounted on a slide using Pro-Texx® mounting medium (Baxter Diagnostics Inc., USA). Microstructure of *sagittae* from $n=327$ post yolk-sac shad larvae, 7.21 to 20.40 mm standard length (mean $L_S \pm$ SD: 12.21 $\pm$ 2.37 mm), was analyzed [vd. 28] using ImagePro Plus® (MediaCybernetics Inc., USA) after processing of the digitized images of the otoliths to improve contrast among *circuli* (Figure 4). The *age* assigned to each larva was the mode of the number of increments counted in three independent readings by the same reader without prior knowledge of length or sample information. The precision of *age* estimates was high: mean coefficient of variation of replicate readings was 5.7%. Moreover, increments width were measured (to the nearest 0.1 μm) in a sub-sample of 30 larvae, randomly chosen from each sampling date.

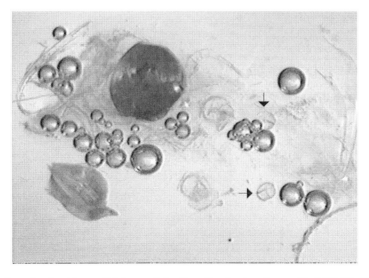

Figure 3. Head of twaite shad larva (16.1 mm L_S) after <3 minutes rinse with 4.5% NaOH. Arrows point to individualized *sagittae*.

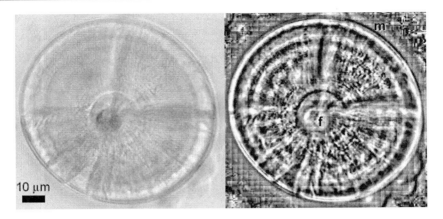

Figure 4. Digitized images of a *sagitta* from a 9.1 mm L_S twaite shad estimated to be 7 days-old (i.e. 7 increments). Image on the left was obtained with a CCD camera mounted on a microscope at 400x magnification. The image was processed using ImagePro Plus® (right) and paired light-and-dark rings were counted from the focus (f) to the margin (m) (see main text for further details).

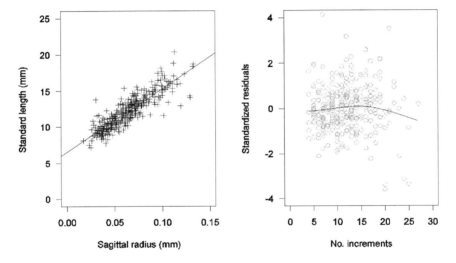

Figure 5. (left) Linear relationship between sagittal radius and larval length of individual Twaite shad larvae (y=6.56+87.56x, n=327, r^2=0.76, p<10^{-4}) of twaite shad caught in River Mira during the sampling season of 2000 (late-March to early-July). (right) Scatter plot of residuals from the previous regression against number of increments counted on *sagittae*. The smooth (spline) curve is shown for illustrative purposes only.

Although no formal validation of ages [*sensu* 29] was attempted here, significant agreement between the number of sagittal rings and the true daily age of larvae has been found for allis shad [12] and American shad [13, 30] indicating that the chronological ages derived from the otoliths are good estimates of the true ages in alosines. Herein, there was a moderate (r^2=0.76) but highly significant (P<10^{-4}) linear relationship between the growth of the otoliths (*sagittae* radius, L_R) and size of larvae ($L_S = 6.56 + 87.56 \cdot L_R$, n=327) (Figure 5), similar to that found for American shad [13, 30], which partially supports the validity of the ageing results and the back calculation method used (see below).

Hatching dates were back-calculated from the date of capture by subtracting the number of increments counted (age) on the otoliths of aged fish and grouped into 7-day classes or cohorts. Hatch date distributions were corrected for cumulative mortality because fish that hatched closer to the time of collection experienced less mortality than older fish in the samples. Since age/size-specific mortality rates of larval twaite shad were unknown *a priori*, we used field, date-specific mortality rates obtained herein (see below). The abundance at age was multiplied by the inverse of the survival rate, $S = e^{-Z \cdot t}$, where $Z \cdot t$ is the cumulative mortality over a period of t days.

GROWTH RATES ESTIMATION

Growth in length of larvae (g, mm d^{-1}) was estimated using:

$$L_t = \alpha + g \cdot t$$

where L_t is the back-calculated L_S (mm) at age t (days) and α is the estimated L_S at hatch (mm). Individual L_S-at-age i were back-calculated considering the body-proportional hypothesis approach [31], $L_t = (c + d \cdot S_i) \cdot (c + d \cdot S_c)^{-1} \cdot L_c$, where c and d are the intercept and slope of the regression of larval L_S on *sagittae* radius, L_i (L_c) and S_i (S_c) are the lengths of larvae and size of *sagittae* (radius) at age i and capture (c) respectively. Backcalculation error was low and no evidence of Lee's Phenomenon was apparent (Figure 6).

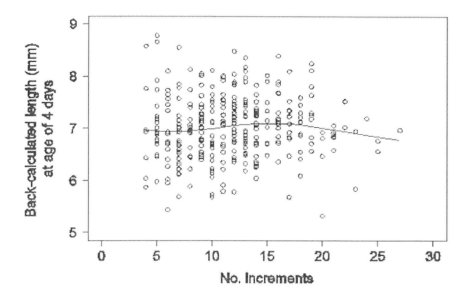

Figure 6. Relationship between individual back-calculated length (mm) at the age of 4 days and the number of increments (age at capture, days) counted on *sagittae* of twaite shad larvae from the River Mira (March-June 2000). The smooth (spline) curve is shown for illustrative purposes only.

Because of the relatively lower number of larvae collected and aged on some dates when compared to other sampling occasions, a model-based resampling (bootstrap) approach was carried out to estimate length-age regression parameters [32, 33]. For each date, the residuals of the regression fit were iteratively resampled and randomly assigned to predicted values of the response-variable. A regression performed on this data was used to obtain estimates of regression coefficients, the size at hatch (intercept, α) and instantaneous growth rate (slope, g^*). From R=9999 iterations or samples, bootstrap averages and standard-errors of coefficients were found and bias-corrected accelerated (BCa) confidence intervals were calculated [34, 35].

ESTIMATION OF MORTALITY RATES

The frequencies of age groups within 0.5 mm length classes, was derived from the sample of $c.$ 330 otolith-aged larvae. This main age-length key (Table I) was used to convert date-specific, corrected length distributions (composed of a total of n=1548 larvae, 4.9 to 20.4 mm L_S) to age distributions from which date-specific mortality rates were then estimated (see below).

Length-frequency data was corrected to account for gear avoidance using $P_C = 0.673 - 0.022 \cdot L_S$ (deduced from Gartz $et\ al.$ [36]) where P_C is the probability of capturing a (striped bass $Morone\ saxatilis$) larva of size L_S (in the 6 – 29 mm range).

Instantaneous daily mortality rates of larvae were estimated from the ln-transformed form of an exponential model of decline in abundance of adjusted numbers with respect to age (time):

$$Z = \left[\ln(N_0) - \ln(N_t)\right]/t$$

where Z is the instantaneous mortality coefficient (d^{-1}), N_0 is the number of larvae (n 100 m^{-3}) of a particular daily cohort ($i.e.$ the group or class of larvae that emerge on a specific day) at time t=0, N_t is the number of larvae (n 100 m^{-3}) of a particular cohort at time t (days after t_0), and t is the time in days (~15 days). The initial abundance (N_i) at time t_i can then be estimated from $\ln(N_i) = \ln(N_0) + Z(t_0 - t_i)$.

Estimates of Z were obtained and compared between sampling dates using analysis of covariance (ANCOVA). When significant, ANCOVA results were studied further using Tukey's simultaneous confidence intervals [37]. Daily mortality percentages were also expressed as $M = (1 - e^{-Z}) \cdot 100\%$ d^{-1}.

Alternatively, the non-linear Pareto model, $\ln(N_i) = \ln(N_0) + \beta t^{\alpha}$, was fitted to data as suggested in Houde [38] to account for age/size-specific changes in mortality. Nonetheless, only in one sampling occasion there was a barely significant ($P<0.1$) increase in the goodness of fit of the non-linear model, thus the more common, simpler log-linear approach described above was used hereafter.

Table I. Age-length key, *i.e.* number of twaite shad larvae of a certain age (no. increments) per 0.5 mm length class

Length-class (mm)	Age (no. increments)																								N
	4	5	6	7	8	9	10	11	12	13	14	15	16	17	18	19	20	21	22	23	24	25	26	27	
(7,7.5]	2																								2
(7.5,8]			1																						1
(8,8.5]	1	2	1	1																					5
(8.5,9]	2	2	3	4	1																				12
(9,9.5]		8	3	7			1	1																	20
(9.5,10]	1	4	4	7	5	3	5																		29
(10,10.5]			2	4	7	5	3	2	1																24
(10.5,11]	1	2	1	2	1	3	4	4		1	1														20
(11,11.5]			2	2	2	7	2	1	3	2	2														23
(11.5,12]		2			1	4	4	3	4	4	3		1	1			1								28
(12,12.5]				1		1	2	6	6	5	1														22
(12.5,13]							2	2	7	2	5	1		2											21
(13,13.5]						2	1	1	7	6	6	2	2	2	1					1					31
(13.5,14]				1				1	1	2	5	3	3		2										18
(14,14.5]								2	1	1	1		2	2		2	1				1	1			14
(14.5,15]										3			3	4	2	1									13
(15,15.5]									1	1		2	1	1	3		1	3							13
(15.5,16]									1	1			1	1	4			1	1						10
(16,16.5]												1	1	1		3		1							7
(16.5,17]																		1	2						3
(17,17.5]															1	3				1					5
(17.5,18]														1											1
(18,18.5]													1											1	2

Table 1. (Continued)

Length-class (mm)	Age (no. increments)																								N
	4	5	6	7	8	9	10	11	12	13	14	15	16	17	18	19	20	21	22	23	24	25	26	27	
(18.5,19]																1					1				2
(19,19.5]																									0
(19.5,20]																									0
(20,20.5]																		1							1
N	7	20	17	29	17	25	24	24	31	28	24	9	16	13	11	10	5	7	3	2	2	2	0	1	327

RELATIONSHIP BETWEEN (A)BIOTIC VARIABLES AND GROWTH AND MORTALITY RATES

Whenever significant correlations were found, least-squares regression analysis were used to assess the relationship among nine environmental and biological variables (Table II) and instantaneous growth and mortality rates of shad larvae.

Table II. List of variables used in this study and summary statistics (mean ± standard deviation (SD) and range) for the period 28 March–7 July 2000

Variables (units)	Mean ± SD	Range
Temperature (°C)	20.58 ± 3.69	16 – 26
Rainfall[§] (mm)	1.59 ± 2.48	0.0 – 8.4
Microplankton biomass (mg AFDW m^{-3})	54.33 ± 36.87	7.51 – 184.8
Crustacean *nauplii* (n 100 m^{-3})	63.41 ± 92.21	0.0 – 396.7
Zooplankton[¶] (n 100 m^{-3})	23.95 ± 36.79	0.0 – 284.3
Predators[†] (n 100 m^{-3})	63.96 ± 453.48	0.0 – 4239.9
Shannon index of diversity H'	0.94 ± 0.43	0.06 – 1.71
Shad larvae (n 100 m^{-3})	26.97 ± 63.19	0.0 – 379.9
Proportion of larvae with gut content	0.22 ± 0.24	0.00 – 0.75

Legend: [§] 7-days moving averages of rainfall; [¶] included isopods, insects, cladocerans and copepodites; [†] included small medusae, mysids and amphipods.

All statistical analyses were conducted using R [39] and a significance level of 0.10 was chosen has a reasonable balance between type I and type II errors, since considerable variation may have been injected by sampling, sub sampling and other measurements.

RESULTS

Twaite shad larvae were subjected to considerable seasonal changes in the environment during their stay in the upper reaches of the River Mira (Figures 7 and 8).

The relatively less abundant shad larvae that emerged earlier in the spawning period (*i.e.* before days 130-140 or mid-May) experienced increased rainfall variability but lower water temperatures as well as lower availability of microplankton and *nauplii*. Conversely, the more abundant daily cohorts that hatched later in the season (*i.e.* after day 140), endured higher temperatures and no precipitation, along with increasing abundances of potential predators/competitors but encountered higher numbers of microplankton and *nauplii* (Figure 8). Despite all sampling efforts, no larvae were collected on day 130, following an unusual, week-long rainfall event (Figure 7), and on day 189 (7 July).

Almost 70% of the *sagittae* analyzed were used (*n*=327) because at least two-out-of-three counts coincided. Shad larvae were aged 4 days to 27 days-old (mean age ± SD: 11.8 ± 4.7 d) but age-range varied among sampling dates/seasonally. The mean width of the first increment deposited on *sagittae* was substantially wider than overall average (6.8 μm *vs.* 5.1 ± 1.05 μm). Afterwards, the increment's average width fluctuated around 5 μm up to the 20th increment and was quite variable for older ages (Figure 9).

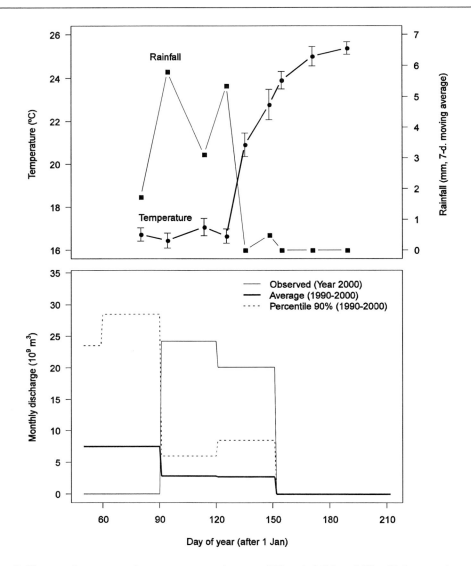

Figure 7. Changes in water surface temperature (mean ± SE), rainfall instability (7-day moving average) and river discharge (monthly data [20]) in River Mira (SW Portugal) with date (day of year after 1 January 2000).

The distribution of cumulative mortality-corrected hatching dates (Figure 10) shows that larval emergence was sparse before early-May (day 130) and peaked during May (days 130 to 150). However, collections of yolk-sac larvae (embryos) and the estimated initial numbers of larvae from catch curves both exhibit a bimodal pattern of hatching; with a first high in March-April (days 90 to 120) and a stronger peak during May (days 130 to 150).

Twaite shad larvae, that hatched at roughly the same size (α^* of 4.7 to 5.9 mm L_s) throughout the season, were estimated to grow at a faster rate, $g^* > 0.5$ mm d^{-1}, earlier in the spawning season when compared to individuals collected after mid-May for which g^* steadily declined from 0.47 mm d^{-1} to 0.33 mm d^{-1} (Table III).

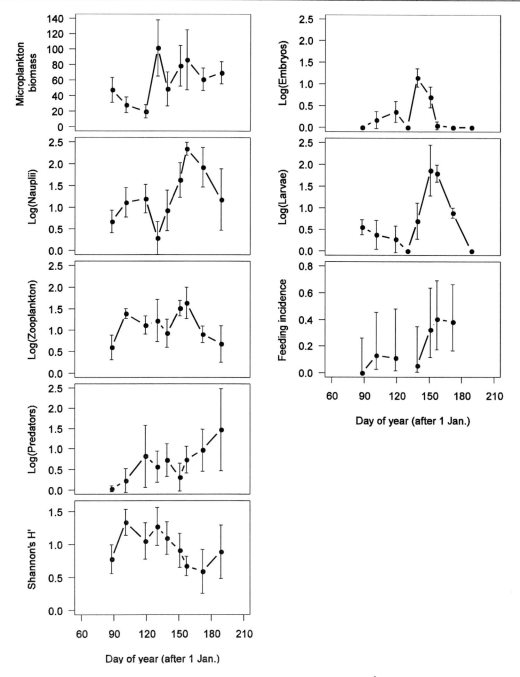

Figure 8. Changes (mean ± SE) in microplankton biomass (mg AFDW m^{-3}), crustacean *nauplii*, zooplankton and predators abundance [log(n 100 m^{-3} +1)], zooplankton diversity (Shannon's *H'*), twaite shad embryos and larvae abundances [log(n 100 m^{-3} +1)] and individuals' feeding incidence (proportion of specimens with gut content) observed in River Mira (SW Portugal) with date (day of year after 1 January 2000).

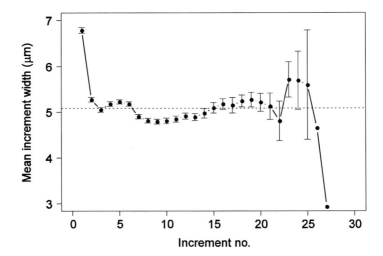

Figure 9. Otolith increment width (mean ± SE) per age-group of twaite shad caught in River Mira (SW Portugal) during the sampling season of year 2000 (late-March to early-July). Horizontal dashed line corresponds to the overall mean (5.1 μm).

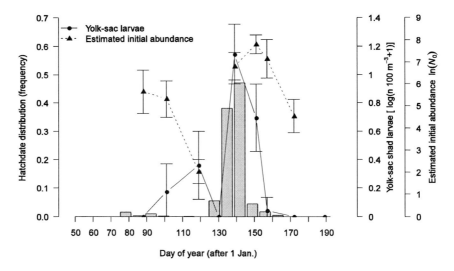

Figure 10. Distribution of the cumulative-corrected hatching dates (bars) vs. the abundance of yolk-sac larvae (–●–). Superimposed is the initial abundance estimated from the catch-curve approach used to determine mortality rates (--▲--) (see main text for further details).

On the other hand, instantaneous mortality rates Z of shad larvae differed significantly among sampling dates (ANCOVA, $F_{[13,52]}=180.3$, $P<10^{-4}$) (Figure 11).

Mortality coefficients were significantly higher (Tukey simultaneous confidence intervals, $P<0.1$) earlier in the spawning season and during a short period in mid-May (-0.486 d^{-1} $<Z<-0.359$ d^{-1} that is 30 to 38.5 % d^{-1}) than later in the season (after day 151: -0.268 d^{-1} $<Z<-0.157$ d^{-1} *i.e.* 23.5 to 14.6 % d^{-1}) (Table IV).

Estimated initial abundances N_0 were significantly ($P<0.01$) lower in late-April (day 119, $N_0=7.5$ n 100 m^{-3}) and higher in late-May (day 151, $N_0=2389.9$ n 100 m^{-3}) than in other sampling occasions (N_0 ranged 92.3 to 1241.4 n 100 m^{-3}). The seasonal pattern of log(N_0) followed the hatch date distribution (Figure 10).

Table III. Date-specific bootstrap estimates (R=9999) of size-at-hatch ($\alpha*$, mm) and instantaneous growth coefficient $g*$ (mm d^{-1}) of twaite shad larvae collected in the River Mira (28 March – 7 July 2000). Day of year is the number of days after 1 January, n_A/n_C is the fraction of aged to collected larvae, size range refers to larvae aged, and 95% CI is the bias-corrected accelerated (BCa) confidence interval

Day of year (Date)	n_A/n_C	Size range (mm)	$\alpha*$ ($\pm$ SE)	95% CI $\alpha*$	$g*$ ($\pm$ SE)	95% CI $g*$
88 (28 March)	6/29	8.4-10.5	4.75 (0.287)	4.18-5.26	0.532 (0.047)	0.446-0.625
101 (10 April)	7/18	10.9-13.2	5.85 (0.088)	5.65-6.00	0.555 (0.016)	0.527-0.588
119 (28 April)	1/9	9.62	NA	NA	NA	NA
130 (9 May)	0/0	NA	NA	NA	NA	NA
139 (18 May)	5/55	8.1-10.8	5.56 (0.666)	4.49-6.55	0.474 (0.131)	0.277-0.474
151 (30 May)	170/769	8.3-15.8	5.50 (0.106)	5.22-5.66	0.417 (0.009)	0.398-0.434
157 (5 June)	127/615	7.2-18.8	5.45 (0.121)	5.18-5.66	0.416 (0.009)	0.396-0.432
172 (20 June)	11/57	11.5-20.4	5.42 (0.099)	5.18-5.59	0.334 (0.009)	0.316-0.350
189 (7 July)	0/0	NA	NA	NA	NA	NA

Legend: NA – not available (no larvae were collected on 9 May and 7 July 2000, days 130 and 189, respectively).

Table IV. Date-specific estimates of cohort initial abundance (ln(N_0), n 100 m^{-3}), instantaneous mortality coefficient Z(d^{-1}) and mortality rate M(% d^{-1}) of twaite shad larvae collected in the River Mira (28 March – 7 June 2000)

Day of year (Date)	Size range (mm)	Age range (d)	Ln(N_0) ($\pm$ SE)	Z (d^{-1}) ($\pm$ SE)	M (% d^{-1})
88 (28 March)	7.4-11.9	7-14	5.632 [a] (0.988)	-0.432 [a] (0.092)	35.1
101 (10 April)	7.1-13.3	9-20	5.297 [a] (0.821)	-0.359 [a] (0.058)	30.2
119 (28 April)	5.2-12.0	4-14	2.021 [b] (0.542)	-0.247 [b] (0.057)	21.9
130 (9 May)	NA	NA	NA	NA	NA
139 (18 May)	4.9-13.0	7-20	6.788 [a] (0.627)	-0.486 [a] (0.050)	38.5
151 (30 May)	6.5-16.3	7-25	7.779 [c] (0.422)	-0.268 [b] (0.025)	23.5
157 (5 June)	7.2-19.2	14-27	7.124 [a] (0.872)	-0.169 [b] (0.043)	15.5
172 (20 June)	6.8-20.4	12-25	4.525 [a] (0.748)	-0.157 [b] (0.039)	14.6
189 (7 July)	NA	NA	NA	NA	NA

Legend: Superscripts a, b, c next to estimated coefficients group similar (i.e. $P>0.1$) estimates column wise. NA – not available (as in Table III).

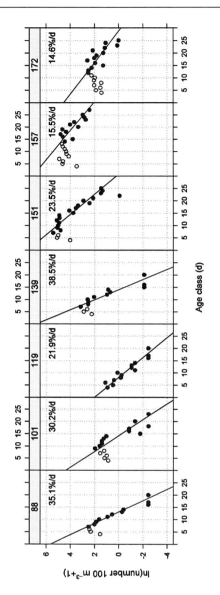

Figure 11. Age-specific mortality curves for twaite shad larvae in the River Mira (SW Portugal) during the year 2000. The abundance estimates for each age-class (circles) were derived from age-length keys. Number above each plot is day of sampling (from 1 January) and inside panels is mortality (% d^{-1}). Age classes used to estimate mortality rates are represented by filled circles. Days of year 101 and 151 corresponded to 10 April and 30 May, respectively (reprinted from Esteves [7] with permission from NovaScience Publishers, Inc.).

Growth and mortality rates were significantly correlated ($P<0.1$) with several environmental variables (Table V). Rates of growth and mortality were also correlated. Stronger associations, herein those with $|r|>0.75$, were found between growth rate and temperature, rainfall, *nauplii* and predators' abundances, and feeding incidence (i.e. the proportion of larvae with gut content), and between mortality rates and *nauplii* abundance and feeding incidence.

Table V. Pearson (product-moment) correlation coefficients for date-specific data (n=6 to 9) collected in the River Mira (28 March – 7 June 2000). Variables and designations as in Table I. Superscripts are used to indicate: NS – non-significant; o – P<0.10, * – P<0.05; ** – P<0.01 and * – P<0.001**

	Growth rate (g*)	M	Temp.	Rainfall	Micropl.	Crustac. nauplii	Zoopl.	Predators	H'	Larvae
Mortality rate (M)	0.79 °									
Temperature	-0.97 **	-0.62 NS								
Rainfall	0.74 °	0.25 NS	-0.83 **							
Microplankton biomass	-0.70 NS	-0.42 NS	0.40 NS	-0.20 NS						
Crustacean *nauplii*	-0.77 °	-0.90 **	0.71 *	-0.58 NS	0.10 NS					
Zooplankters	-0.15 NS	-0.49 NS	-0.01 NS	0.23 NS	0.27 NS	0.45 NS				
Predators	-0.79 °	-0.59 NS	0.66 °	-0.44 NS	0.15 NS	0.27 NS	-0.27 NS			
Shannon's H'	0.73 NS	0.55 NS	-0.68 *	0.80 **	-0.21 NS	-0.67 *	0.20 NS	-0.26 NS		
Shad larvae	-0.57 °	-0.46 NS	0.46 NS	-0.52 NS	0.29 NS	0.75 *	0.60 °	-0.25 NS	-0.52 NS	
Feeding incidence	-0.83 *	-0.88 **	0.85 *	-0.45 NS	0.74 °	0.97 ***	0.61 NS	0.46 NS	-0.58 NS	0.76 *

Moreover, rainfall and temperature (r=-0.83), temperature and feeding incidence (r=+0.85), Shannon's *H'* and rainfall instability (r=+0.80), and crustacean *nauplii* abundance and feeding incidence (r=+0.97) were significantly correlated (*P*<0.01). Regression models describing the most relevant relationships between growth and mortality rates and covariates are presented in Table VI.

Tablr VI. Regression models obtained for growth rate (*g**, mm d⁻¹) and mortality (*M*, % d⁻¹) of twaite shad larvae collected in the River Mira (28 March – 7 June 2000)

Trait	Variable	n	Intercept (± SE)	Slope (± SE)	R^2	F (p-value)
Growth (g*)	Temperature	6	0.911 (0.062)	-0.022 (0.003)	0.933	55.88 (0.0017)
	Rainfall	6	0.419 (0.030)	+0.027 (0.092)	0.548	4.85 (0.0924)
	Crustacean *nauplii*	6	0.597 (0.064)	-0.099 (0.041)	0.591	5.79 (0.0739)
	Predators	6	0.545 (0.042)	-0.180 (0.070)	0.625	6.67 (0.0612)
	Feeding incidence	6	0.538 (0.035)	-0.386 (0.131)	0.683	8.62 (0.0452)
Mortality (*M*)	Crustacean *nauplii*	7	45.19 (4.84)	-13.99 (3.22)	0.791	18.92 (0.0074)
	Feeding incidence	7	35.46 (3.05)	-49.07 (12.06)	0.768	16.55 (0.0097)

DISCUSSION

Twaite shad larvae in the River Mira do not grow at the same rate or experience uniform mortality during the spawning season. Differences in the relative temporal pattern of growth and mortality rates were observed for early and late daily-cohorts of shad larvae. The two groups were subjected to comparable and declining mortality rates but the early-spawned, less abundant larvae (collected before May, *c.* day 130) grew at a significantly faster rate (*g**>0.5 mm d⁻¹) than those caught later in the season (from day 139 to day 172) for which *g** decreased from 0.47 to 0.33 mm d⁻¹.

Date-specific growth rates *g** are similar to those reported by Crecco and Savoy [13] for American shad, 0.38 to 0.52 mm d⁻¹ but are greater than the growth rate that can be estimated from age-at-length data of reared *A. alosa* obtained by Lochet *et al.* [12], 0.269 mm d⁻¹. The difference between fast- and slow-growing cohorts was 0.22 mm d⁻¹. Over a 20 days period this could mean a 4.4 mm difference in larval length or, alternatively, this would represent an 18 days difference in the duration of the larval stage, assuming that metamorphosis into juveniles, *i.e.* the "transitional stage", occurs at ≥20 mm [10]. The shorter the stage duration the shorter the period of time larvae are exposed to episodic mortalities associated with extreme weather events which are significant to freshwater fish (see below) and the temporal window within which vital rates may vary to affect survivorship [40]. On the other hand, estimated length-at-age using the models provided herein compare well with sizes at age referred to by Quignard and Douchement [10]: 5-8 mm at hatch, 8-9 mm at 6 days-old and 13-15 mm for 15-20 days-old twaite shad larvae. Moreover, our estimates of size-at-hatch (*c.* 5-6 mm L_S) correspond satisfactorily to published data from field and rearing studies of twaite shad [*cf.* 41].

Several environmental variables were related to the seasonal changes in the growth of twaite shad larvae in the River Mira. The early-spawned shad larvae seemed to grow faster when a unique, almost irreconcilable set of environmental conditions were found: lower temperatures; highly unstable rainfall conditions; lower microplankton biomass and reduced *nauplii* abundances but higher plankton diversity. In addition, a smaller fraction of those individuals had visible gut content (our index of feeding incidence). Unexpectedly, growth was not positively related to temperature and/or prey availability/food ingestion [6] as found before for American shad [13, 18] and threadfin shad *Dorosoma petenense* (Gunther, 1867) [42]. Eventually, larval size and growth are not the only traits that need to be considered. The work of Lee Fuiman and James Cowan [43] indicate that larvae having a suite of physiological or behavioral characteristics that confer greater fitness in predator-prey situations seem to exist. Possibly we sampled and analyzed these larval "athletes" that survived the theoretically poorer conditions found earlier in the spawning season, thence being able to feed, grow and avoid predators. Seemingly, a large fraction of the yolk-sac larvae found in earlier samples did not survive to be aged on later collections. In a complementary study of twaite shad larvae in the upper reaches of River Mira, Esteves et al. [21] found that RNA/DNA ratios – an ecophysiological index of condition, susceptible to environmental changes, that has been linked to survival and growth both in laboratory and field studies – were fairly constant throughout the period studied (March-June 2000). In addition, particular environmental conditions, namely water temperatures of *c.* 22 °C and freshwater pulses [*sensu* 44] (resulting from the changing rainfall, a proxy of river flow), largely contributed to enhance the RNA/DNA ratio [21].

Herein, the widths of increments deposited on *sagittae* of twaite shad larvae (1.8 to 12.2 μm), were well above the practical resolution limit of light microscopy, $0.5 - 1.0$ μm [45, 46], and within the range observed by Limburg [16] for American shad collected in the Hudson river estuary (USA), 2 to 7 μm for larvae 9 to 27 mm SL, but were substantially narrower than the increments measured by Savoy and Crecco [30] on the *sagittae* of American shad captured in the Connecticut River (USA) (13.5 to 24.6 microns for larvae 4 to 24 days-old). Furthermore, average increment widths measured in this study exhibited a rather convoluted pattern with increasing age, being higher than average for the first, *c.* 5th and 15-20th increments. These correspond roughly to milestones in shad ontogeny namely hatch (in fact, the otoliths are visible immediately after hatch [47]), yolk-sac resorption and end of larval period [10, 47] respectively and may reflect the improved ontogenic conditions for larval (and otolith) growth at these transition periods (end of embryonic development, completion of yolk-sac resorption and increased swimming and feeding capabilities, respectively).

The natural changes in the seasonal distribution of twaite shad larvae and thence the high variation in the number of larvae caught and aged on each sampling date strongly affected the estimates of growth rates despite the care taken in the ageing procedure and the use of alternative, computationally intensive procedures (bootstrap). This source of error could have unduly interfered with the unexpected relationships found between growth rates and environmental covariates.

Date-specific mortality rates of twaite shad larvae declined steadily from 35 to 22% d^{-1} during a 31-day interval between late-March and late-April and again from 39 to 15% d^{-1} in a second period of 33 days later in the spawning season (mid-May to late-June). This pattern of seasonal variation in mortality rates differs from the results of Crecco and Savoy [14] for American shad in the Connecticut River (USA). Using a different methodology to estimate mortalities, these authors found that cohort-specific mortality rates were higher for early-spawned cohorts (c. 6 to 7% d^{-1}) and declined gradually to about 2 – 3% d^{-1} later in the spawning season. The two "periods" observed herein are possibly a consequence of the unexpected high rainfall events and the resulting high river flows registered in the River Mira during late-April and early-May. On one hand, Leach and Houde [18] demonstrated that environmental pulses of temperature and pH, e.g. 5 °C rise in temperature or reductions of pH from 7.0 to 6.0 that may occur in association with extreme rainfall events, had strong detrimental responses in survival probability of reared American shad larvae. In addition, the magnitude of larval American shad mortality (and juvenile production) in the Connecticut River (USA) among discrete cohorts depends largely on the coupling of larval emergence and river flow conditions one week later [14]. On the other hand, those extreme hydrological events may have advected larvae and their zooplankton prey some distance from "safe" sites, where conditions are favorable, to nearby sections of the river of lower food density, high predator abundance [14] or even downriver towards the euryhaline estuary. Therein, high or complete mortality is expected to be induced by direct seawater exposure of unprepared individuals [48, 49, 50]. Moreover, the pattern of larval shad abundance might reflect a bimodal pattern of spawning runs [51] and/or the repeated/serial spawning behavior of shad [52, 53, 54], during which similar mortality patterns were observed. Our data does not clarify what occurred.

Houde [38] reviewed the several causes of early life mortality, viz. starvation and nutritional deficiencies, predation, physics (transport, retention and dispersal), water quality and nursery habitats, diseases and habitats and their interactions. Our results of correlation analysis suggest that crustacean nauplii abundance and feeding incidence are important variables affecting mortality of twaite shad larvae. In fact, survival of fish larvae has been linked to starvation namely in Atlantic herring and Atlantic cod [38]. Short periods of food deprivation among (first-feeding) larvae can result in death from malnutrition [55, 56] because the foraging ability of young herring-like larvae is greatly limited by their small mouth and poor swimming ability [56]. The strong coupling between M and nauplii abundance and larvae feeding incidence favors the hypothesis that mortality of twaite shad larvae is (directly) controlled via a bottom-up mechanism that contrasts with the point of view currently accepted that predation is the major source of larval mortality [38]. Nonetheless, it remains problematic or impossible to quantify the proportional mortalities from starvation or predation in natural ecosystems despite the intuitiveness of the idea that poorly fed, slowly growing larvae are more vulnerable to predators [38] – an alternative expression of the Bigger-is-Better-Hypothesis [43]. Herein, we did not find evidence of the direct impact of predation/competition upon mortality rate but, in contrast, growth rate $g*$ was inversely correlated with the abundance of predators/competitors.

CONCLUSION

Twaite shad larvae in the River Mira (SW Portugal) did not grow at the same rate or experience uniform mortality during the spawning season. Early and late daily-cohorts of shad larvae were subjected to comparable and declining mortality rates (from 35 to 22% d^{-1} during a 31-day interval between late-March and late-April and again from 39 to 15% d^{-1} in a second period of 33 days later in the spawning season – mid-May to late-June) but the early-spawned, less abundant larvae (collected before May) grew at a significantly faster rate ($g^*>0.5$ mm d^{-1}) than those caught later in the season (from mid-May to mid-June) for which g^* decreased from 0.47 to 0.33 mm d^{-1}.

Changes in the abundance of potential prey, namely crustacean *nauplii*, coincident with early larval development (essential in the ability of capturing, handling and consuming the prey successfully), are important regulators of mortality of twaite shad in the upper stretches of the River Mira. In contrast, we found seemingly conflicting effects of the studied (a)biotic factors upon the seasonally changing growth rates. The ageing procedure and the reduced number of larvae aged in some occasions, due to the natural, seasonal distribution pattern of the specimens might explain the results obtained for growth rate of twaite shad larvae. The extension of the analysis described herein to a longer period, encompassing several spawning seasons of twaite shad in the River Mira, and the carrying out of complementary experimental work under controlled conditions [*cf.* 18, 19] would most certainly clarify some aspects of our results. Despite recent contributions to improve knowledge of twaite shads' early life-history stages, particularly of European species [see *e.g.* 7], work is still necessary to address yet unanswered questions dealing with the fundamental aspects of biology, traits such as growth and mortality, in less-studied species, *e.g.* twaite shad, and Mediterranean and Ponto-caspian species

ACKNOWLEDGMENTS

We gratefully acknowledge funding by the Fundacão para a Ciência e Tecnologia (via a research project, ref. PRAXIS XXI/CA/3/3.2/1981/95, and a grant to Eduardo Esteves, ref. PRAXIS XXI/BD/18206/98). Teresa Pina, Jorge Palma, João Quintela and José António Bentes helped extensively in the field work.

REFERENCES

[1]   Cabral, M. J., Queiroz, A. I., Palmeirim, J., Almeida, J., Rogado, L., Santos-Reis, M., Oliveira, M. E., Ferrand Almeida, N., Almeida, P. R. and Dellinger, T. (2006). *Livro Vermelho dos Vertebrados de Portugal*. Lisboa: Instituto de Conservação da Natureza.

[2]   Alexandrino, P., Faria, R., Linhares, D., Le Corre, M., Sabatié, R., Baglinière, J. L. and Weiss, S. (2006). Interspecific differentiation and intraspecific substructure in two closely related clupeids with extensive hybridization, *Alosa alosa* and *Alosa fallax*. *Journal of Fish Biology, 69(Suppl. B)*, 242-259.

[3]   Waldman, J. R. (2003). Introduction to the shads. *American Fisheries Society Symposium 35*, 3-9.

[4]   Assis, C. A. (1990). Threats to the survival of anadromous fishes in the River Tagus, Portugal. *Journal of Fish Biology, 37(Suppl. A)*, 225-226.

[5]   Houde, E. D. (1987). Fish early life dynamics and recruitment variability. *American Fisheries Society Symposium, 2*, 17-29.

[6]   Houde, E. D. (1996). Evaluating stage-specific survival during the early life of fish. In Watanabe, Y., Yamashita, Y. and Ooseki, Y. (Eds.), *Survival strategies in early life stages of marine resources* (pp. 51-66) Roterdam: A.A. Balkema.

[7]   Esteves, E. (2011). Ecology of early life-history stages of anadromous shads. In Dempsey, S. P. (Ed.), *Fish Ecology* (pp. 151-172) New York: Nova Science Publishers, Inc.

[8]   Aprahamian, M. W., Baglinière, J.-L., Sabatié, M. R., Alexandrino, P., Thiel, R. and Aprahamian, C. D. (2003). Biology, status, and conservation of the anadromous Atlantic Twaite shad *Alosa fallax fallax*. *American Fisheries Society Symposium, 35*, 103-124.

[9]   Baglinière, J.-L. and Elie, P. (2000). *Les aloses (Alosa alosa et Alosa fallax spp.)*. Paris: Cemagref Editions et INRA Editions.

[10]   Quignard, J. P. and Douchement, C. (1991). *Alosa fallax fallax* (Lacepède, 1803). In Hoestland, H. (Ed.), *The freshwater fishes of Europe. Vol. 2. Clupeidae, Anguillidae* (pp. 225-253) Wiesbaden: AULA-Verlag Wiesbaden.

[11]   Lochet, A., Boutry, S. and Rochard, E. (2009). Estuarine phase during seaward migration for allis shad *Alosa alosa* and twaite shad *Alosa fallax* future spawners. *Ecology of Freshwater Fish, 18*, 323-335.

[12]   Lochet, A., Jatteau, P., Tomás, J. and Rochard, E. (2008). Retrospective approach to investigating the early life history of a diadromous fish: allis shad *Alosa alosa* (L.) in the Gironde–Garonne–Dordogne watershed. *Journal of Fish Biology, 72*, 946-960.

[13]   Crecco, V. and Savoy, T. (1985). Effects of biotic and abiotic factors on growth and relative survival of young American shad, *Alosa sapidissima*, in the Connecticut River. *Canadian Journal of Fisheries and Aquatic Sciences, 42*, 1640-1648.

[14]   Crecco, V. and Savoy, T. (1987). Effects of climatic and density-dependent factors on intra-annual mortality of larval American shad. *American Fisheries Society Symposium, 2*, 69-81.

[15]   Crecco, V. and Savoy, T. (1987). Review of recruitment mechanisms of the American shad: the critical period and match-mismatch hypothesis reexamined. *American Fisheries Society Symposium, 1*, 455-468.

[16]   Limburg, K. E. (1996). Growth and migration of 0-year American shad (*Alosa sapidissima*) in the Hudson River estuary: otolith microstructural analysis. *Canadian Journal of Fisheries and Aquatic Sciences, 53*, 220-238.

[17]   Limburg, K. E. (1996). Modelling the ecological constraints on growth and movement of juvenile American shad (*Alosa sapidissima*) in the Hudson River Estuary. *Estuaries, 19*, 794-813.

[18]   Leach, S. D. and Houde, E. D. (1999). Effects of environmental factors on survival, growth, and production of American shad larvae. *Journal of Fish Biology, 54*, 767-786.

[19]   Letcher, B. H., Rice, J. A., Crowder, L. B. and Binkowski, F. P. (1997). Size- and species-dependent variability in consumption and growth rates of larvae and juveniles

of three freshwater fishes. *Canadian Journal of Fisheries and Aquatic Sciences*, *54*, 405-414.

[20] INAG. (1995-2011). "Sistema Nacional de Informação de Recursos Hídricos (SNIRH)." Instituto Nacional da Água, Lisboa, Portugal. Retrieved 20-10-2011, from http://snirh.inag.pt.

[21] Esteves, E., Pina, T. and Andrade, J. P. (2009). Diel and seasonal changes in nutritional condition of the anadromous Twaite shad *Alosa fallax fallax* (Lacépède, 1803) larvae *Ecology of Freshwater Fish*, *18*, 132-144.

[22] Smith, D. L. (1977). *A guide to marine and coastal plankton and marine invertebrate larvae*. Dubuque, Iowa: Kendall/Hunt Company.

[23] Todd, C. D. and Laverack, M. S. (1991). *Coastal marine zooplankton. A pratical manual for students*. Cambridge: Cambridge University Press.

[24] Allan, J. D. (1995). *Stream ecology. Structure and functions of running waters*. London: Chapman and Hall.

[25] Newell, G. E. and Newell, R. C. (1963). *Marine plankton, a pratical guide*. London: Hutchinson of London.

[26] Goldman, C. R. and Horne, A. J. (1983). *Limnology*. Auckland: McGraw-Hill.

[27] Krebs, C. J. (1989). *Ecological methodology*. New York: Harper Collins Publishers Inc.

[28] Secor, D. H., Dean, J. M. and Laban, E. H. (1992). Otolith removal and preparation for microstructural examination. In Stevenson, D. K. and Campana, S. E. (Eds.), *Otolith microstructure examination and analysis* (pp. 19-57) Ottawa: Canadian Special Publications on Fisheries and Aquatic Sciences.

[29] Geffen, A. J. (1992). Validation of otolith increment deposition rate. In Stevenson, D. K. and Campana, S. E. (Eds.), *Otolith microstructure examination and analysis* (pp. 101-113) Ottawa: Canadian Special Publications on Fisheries and Aquatic Sciences.

[30] Savoy, T. and Crecco, V. (1987). Daily increments on the otoliths of larval American shad and their potential use in population dynamics studies. In Summerfelt, R. C. and Hall, G. E. (Eds.), *The age and growth of fish* (pp. 413-431) Ames, Iowa: The Iowa State University Press.

[31] Francis, R. I. C. C. (1990). Back-calculation of fish length: a critical review. *Journal of Fish Biology*, *36*, 883-902.

[32] Crawley, M. J. (2007). *The R Book*. Chichester, England: Wiley Publishing.

[33] Draper, N. R. and Smith, H. (1998). *Applied Regression Analysis* (3rd edition). New York: John Wiley and Sons Inc.

[34] Canty, A. and Ripley, B. boot: Bootstrap R (S-Plus) Functions. R package version 1.2-36. [Online]. 2009 Available from http://cran.r-project.org/web/packages/boot/index.html

[35] Davison, A. C. and Hinkley, D. V. (1997). *Bootstrap methods and their applications*. Cambridge, UK: Cambridge University Press.

[36] Gartz, R. G., Miller, L. W., Fujimura, R. W. and Smith, P. E. (1999). Measurement of larval striped bass (*Morone saxatilis*) net avoidance using evasion radius estimation to improve estimates of abundance and mortality. *Journal of Plankton Research*, *21*, 561-580.

[37] Venables, W. N. and Ripley, B. D. (2002). *Modern applied statistics with S-Plus* (4th edition). New York: Springer-Verlag.

[38] Houde, E. D. (2002). Mortality. In Fuiman, L. A. and Werner, R. G. (Eds.), *Fishery Science. The unique contributions of early life stages* (pp. 64-87) Oxford, UK: Blackwell Science/Publishing.

[39] R Development Core Team (2007). *R: A language and environment for statistical computing*. Vienna, Austria: R Foundation for Statistical Computing.

[40] Houde, E. D. (1994). Differences between marine and freshwater fish larvae: implications for recruitment. *ICES Journal of Marine Science, 51*, 91-97.

[41] Taverny, C., Cassou-Leins, J. J., Cassou-Leins, F. and Elie, P. (2000). De l'ouef à l'adulte en mer. In Baglinière, J.-L. and Elie, P. (Eds.), *Les aloses (Alosa alosa et Alosa fallax spp.)* (pp. 93-124) Paris: INRA Editions et Cemagref Editions.

[42] Betsill, R. K. and Avyle, M. J. v. d. (1997). Effect of temperature and zooplankton abundance on growth and survival of larval threadfin shad. *Transactions of the American Fisheries Society, 126*, 999-1011.

[43] Cowan Jr., J. H. and Shaw, R. F. (2002). Recruitment. In Fuiman, L. A. and Werner, R. G. (Eds.), *Fishery Science. The unique contributions of early life stages* (pp. 88-111) Oxford, UK: Blackwell Science/Publishing.

[44] Chícharo, L. M., Chícharo, M. A. and Ben-Hamadou, R. (2006). Use of hydrotechnical infrastructure (Alqueva Dam) to regulate planktonic assemblages in the Guadiana estuary: Basis for sustainable water and ecosystem services management. *Estuarine, Coastal and Shelf Science, 70*, 3-18.

[45] Campana, S. E. (1992). Measurement and interpretation of the microstructure of fish otoliths. In Stevenson, D. K. and Campana, S. E. (Eds.), *Otolith microstructure examination and analysis* (pp. 59-71) Ottawa: Canadian Special Publications on Fisheries and Aquatic Sciences.

[46] Neilson, J. D. (1992). Sources of error in otolith microstructure examination. In Stevenson, D. K. and Campana, S. E. (Eds.), *Otolith microstructure examination and analysis* (pp. 115-126) Ottawa: Canadian Special Publications on Fisheries and Aquatic Sciences.

[47] Taverny, C., Cassou-Leins, J. J., Cassou-Leins, F. and Elie, P. (2000). De l'oeuf à l'adulte en mer. In Baglinière, J.-L. and Elie, P. (Eds.), *Les aloses (Alosa alosa et Alosa fallax spp.)* (pp. 93-124) Paris: Cemagref Editions et INRA Editions.

[48] Bardonnet, A. and Jatteau, P. (2008). Salinity tolerance in young Allis shad larvae (*Alosa alosa* L.). *Ecology of Freshwater Fish, 17*, 193-197.

[49] Leguen, I., Veron, V., Sevellec, C., Azam, D., Sabatie, R., Prunet, P. and Bagliniere, J. L. (2007). Development of hypoosmoregulatory ability in allis shad Alosa alosa. *Journal of Fish Biology, 70*, 630-637.

[50] Zydlewski, J. and McCormick, S. D. (1997). The ontogeny of salinity tolerance in the American shad, *Alosa sapidissima. Canadian Journal of Fisheries and Aquatic Sciences, 54*, 182-189.

[51] Mennesson-Boisneau, C., Aprahamian, C. D., Sabatié, M. R. and Cassou-Leins, J. J. (2000). Remontée migratoire des adultes. In Baglinière, J.-L. and Elie, P. (Eds.), *Les aloses (Alosa alosa et Alosa fallax spp.)* (pp. 55-72) Paris: INRA Editions et Cemagref Editions.

[52] Pina, T. (2000). Reprodução de savelha *Alosa fallax fallax* (Lacépède, 1803) nos rios Mira e Guadiana. (MSc Dissertation) Faculdade de Ciências do Mar e Ambiente, Universidade do Algarve, Faro, Portugal.

[53] Pina, T., Esteves, E. and Andrade, J. P. (2003). Gross and histological observations of ovarian development in twaite shad, *Alosa fallax fallax*, from the rivers Mira and Guadiana (Portugal). *Scientia Marina, 67*, 313-322.

[54] Aprahamian, M. W. (1982). Aspects of the biology of the twaite shad, *Alosa fallax fallax* (Lacépède), in the Rivers Severn and Wye. (Doctoral Thesis) University of Liverpool, UK.

[55] Blaxter, J. H. S. and Hunter, J. R. (1982). The biology of clupeoid fishes. *Marine Biology, 20*, 1-223.

[56] Hunter, J. R. (1981). Feeding ecology and predation of marine fish larvae. In Lasker, R. (Ed.), *Marine fish larvae. Morphology, ecology and relation to fisheries* (pp. 33-77) Washington: Washington Sea Grant Program.

In: Larvae: Morphology, Biology and Life Cycle
Editors: Kia Pourali and Vafa Niroomand Raad

ISBN: 978-1-61942-662-7
© 2012 Nova Science Publishers, Inc.

Chapter 5

VARIATION IN IMMUNOGENICITY AND INFECTIVITY OF ENCAPSULATING AND NON-ENCAPSULATING *TRICHINELLA* SPECIES

Emília Dvorožňáková and Zuzana Hurníková*
Institute of Parasitology of the Slovak Academy of Sciences,
Košice, Slovak Republic

ABSTRACT

Trichinellosis is a zoonotic disease with worldwide distribution caused by intestine nematode of the genus *Trichinella*. Infectivity of this parasite is determined by host immune and inflammatory response to the infection. The murine cellular and humoral immune response to the infection with low doses of larvae of encapsulating (*Trichinella spiralis, T. britovi*) and non-encapsulating (*T. pseudospiralis*) species was studied. Mice were experimentally infected with 10 muscle larvae of the parasite to simulate natural conditions of the infection in rodents, important reservoirs of trichinellosis.

Both *T. spiralis* and *T. britovi* stimulated the proliferation of splenic T and B lymphocytes during the intestinal phase of infection, but *T. spiralis* activated the proliferative response also at the muscle phase, particularly in B cells. Non-encapsulating *T. pseudospiralis* stimulated the proliferation of T and B cells only on day 10 post infection (p.i.) and later at the muscle phase. The numbers of splenic CD4 and CD8 T cells of *T. spiralis* infected mice were significantly increased till day 10 p.i., i.e. at the intestinal phase, and then at the late muscle phase, on day 60 p.i. *T. britovi* infection increased the CD4 and CD8 T cell numbers only on day 30 p.i. Decreased numbers of CD4 and CD8 T cells after *T. pseudospiralis* infection suggest a suppression of cellular immunity. Both encapsulating *Trichinella* species induced the Th2 response (cytokines IL-5, IL-10) at the intestinal phase and the Th2 dominant response at the advanced muscle phase. IFN-γ production (Th1 type) started to increase with migrating newborn larvae from day 15 p.i. till the end of the experiment. IL-5 production was suppressed during the intestinal phase of *T. pseudospiralis* infection. The immune response to *T.*

---

* Institute of Parasitology of the Slovak Academy of Sciences, Hlinkova 3, 040 01 Košice, Slovak Republic, e-mail: dvoroz@saske.sk.

pseudospiralis was directed more to the Th1 response at the muscle phase, the high IFN-γ production was found on day 10 p.i. and it peaked on days 45 and 60 p.i. The low infective dose of *T. spiralis*, *T. britovi*, and *T. pseudospiralis* induced a late seroconversion in infected mice. Both *T. spiralis* and *T. britovi* did not evoke an increased specific IgM response, which is typical for the acute infection. Only *T. pseudospiralis* induced a higher specific IgM level in the intestinal phase of the infection, till day 30 p.i. *T. spiralis* caused earlier and more intensive specific antibody response (IgG$_1$, IgG$_{2a}$, IgG$_{2b}$), from day 45 p.i, when antigens from newborn and muscle larvae were accumulated, on the contrary to *T. britovi* (IgG$_1$, IgG$_{2a}$, IgG$_{2b}$) and *T. pseudospiralis* (IgG$_{2b}$), which induced specific antibody response from day 60 p.i.

Knowledge of the functional consequences of species species-specific parasite infection may be used in immunodiagnosis of trichinellosis as well as immunologist immunologic strategies to control the infection.

Keywords: *Trichinella* spp., T lymphocytes, Th1/Th2 immune response, cytokines

INTRODUCTION

Trichinella spp. is the intestinal nematode parasite with worldwide distribution and which causes trichinellosis – a serious zoonosis. (Miterpáková et al. 2009; Paraličová et al. 2009). At present, the genus *Trichinella* comprises five encapsulating (*T. spiralis*, *T. nativa*, *T. britovi*, *T. nelsoni* and *T. murrelli*) and three non-encapsulating species (*T. pseudospiralis*, *T. papuae* and *T. zimbabwensis*)(Morgan 2000). The characteristics of these species are based on biological, biochemical and genetic criteria. Species variation in infectivity and immunogenicity is very important among biological parameters (Bolas-Fernández 2003). Experimental studies demonstrated significant differences in infectivity of well characterized domestic and sylvatic *Trichinella* genotypes and the dependence of infective dose on the effectiveness of infection (Kapel and Gamble 2000; Šnábel et al. 2001; Cui et al. 2006; Reiterová et al. 2009). Infectivity of *Trichinella* species is also determined by the immune status of the host, host's immune and inflammatory responses to species-specific antigens (Bolas-Fernández 2003). Genetically determined variation in species of genus *Trichinella* and in host's response play an important role in the outcome of infection and are studied in details. However, much less is known of its influence on the host-parasite relationship. Experimental studies on mice can demonstrate how different *Trichinella* species influence the nature and degree of the host immune and inflammatory responses, the complex interplay between immunogenicity and pathogenicity influencing both host and parasite, how species-specific parasite affects host capacity to develop the Th cell response necessary for resistance, cytokine production and kinetics of specific antibodies (Wakelin et al. 2002; Bolas-Fernández 2003).

The complete life cycle of *Trichinella* spp. occurs in one host, which means that the host can exert its immune attack at several points in the life cycle. Adult worms reproduce in the small intestine, new born larvae migrate through the blood and lymphatics to the skeletal muscle cells, and encapsulated or non-encapsulated larvae (L1) represent the infective stage of the parasite (Despommier 1998). Every stage of the life cycle of trichinella can evoke a stage-specific protective host immune response due to the uniqueness in both the cuticular and the excretion/secretion antigens of each stage (Bruschi et al. 1992;, Wang 1997). The mechanism of expulsion of worms is dependent on the Th2 type of response, what leads to the activation of mucosal mast cells (Khan et al. 2001). CD4+ T helper type 2 cells are critical in host protective immune and inflammatory responses during trichinella intestinal infection. Muscle infection elicites a chronic infection where a major role in host defense processes is played by cellular immunity (Mahida 2003). Larvae survive in nurse cells in close association with macrophages, CD8+ and CD4+ T lymphocytes, and B lymphocytes. B lymphocytes secreting antibodies, particularly IgG and IgE, may lead to an effective antibody-dependent cell mediated cytotoxic reaction against newborn larvae (Beiting et al. 2004).

The earlier specific antibodies are bound to *Trichinella* antigens and form immune complexes, which are present in infected host at the beginning of the infection (Dziemian & Machnicka 2000; Feldmeier et al. 1987). The protective isotypes IgG_1 and IgG_2 are involved in the inflammatory response. An elevation of IgG_1 accompanies the muscle phase of infection (Doligalska 2000) and newborn larve are more sensitive *in vitro* to IgG_1 in antibody-dependent cellular cytotoxicity (Moskwa 1999). Antibodies significantly participate in trichinella entrapment and rapid expulsion of infective larvae L1, reduce adult worm fecundity and kill newborn larvae (Appleton and & Usack 1993). The consistent release of circulating antigens by the larvae plays a major role in sustaining the host immune response until the calcification of the parasites (Li et al. 1999). Circulating antigens are present in plasma and urine of infected organisms about 30 days after the infection (Machnicka et al. 2001, ; Kolodziej-Sobocińska et al. 2006). Specific antibodies remain detectable for a very long time after the infection. Polyclonal lymphocyte activation of T-cells, but particularly B-cells, is responsible for the high levels of immunoglobulines IgG, IgM, and IgE observed in infected animals and humans (Murrel & Bruschi, 1994).

Specific anti-*Trichinella* IgM antibodies are first found, during second week of the infection (Li & Ko 2001, ; Kolodziej-Sobocińska et al. 2006). Specific IgG antibodies were found also 6 or 8 months after infection, even after 3 years (Dziemian a & Machnicka 2000, ; Kolodziej-Sobocińska et al. 2006, ; Morales et al. 2002). Stimulation of ES excretory-secretory antigens production by larvae and their penetration through the capsules (Pritchard, 1985) as well as degradation of larvae by inflammatory cells (Candolfi et al., 1989) explains the long-lasting presence of IgM, IgG antibodies. Immunoglobulines IgG_1 represent Th2-cell activation and IgG_2 antibodies reflect Th1 response (Else and & Finkelman, 1998). A significant elevation of IgG_1 is often observed in trichinellosis (Li & Ko 2001, ; Kolodziej-Sobocińska et al. 2006). Both IgG_1 and IgG_2 are responsible for antilarval activity of peritoneal eosinophils, that are involved in the inflammatory response (Doligalska, 2000).

Potential differences in a development of the host immune response to different *Trichinella* species and host-parasite interactions are rarely studied under *in vivo* conditions (Malakauskas et al. 2001; Bolas-Fernández 2003; Lee and Ko 2006). Kapel and Gamble (2000) observed variances in the infectivity and antibody responses of pigs to domestic and sylvatic *Trichinella* spp. after a high infective dose. Andrade et al. (2007) described interspecies results from *in vitro* study, where the differences in NO production of macrophages after stimulation with L1 antigens from encapsulated encapsulating and non-encapsulated encapsulating trichinellas were demonstrated. Immunochemical variety of larval *T. spiralis* and *T. pseudospiralis* antigens have been confirmed in several studies (Turčeková et al. 1997; Ros-Moreno et al. 2002; Robinson et al. 2007).

The aim of this study was to compare infectivity and immunogenicity of encapsulating (*Trichinella spiralis*, *T. britovi*) and non-encapsulating (*T. pseudospiralis*) species in mice after the infection with low doses of species-specific larvae.

MATERIALS AND METHODS

The experiment was carried out on male BALB/c mice (n=144) weighting 20-25 g. Mice were kept under a 12-h light/dark regime at room temperature (22-24°C) and 56 % humidity on a commercial diet and water. The experimental protocol was approved by the Institute of Parasitology Animal Care Commitee. Animals were divided randomly into four groups as folows: Group 1 (n=24) was uninfected and untreated (control), mice in Group 2 (n=40) were infected *per os* with 10 *T. spiralis* larvae per mouse on day 0 of the experiment. Mice in Groups 3 (n=40) and 4 (n=40) were infected *per os* with 10 *T. britovi* and *T. pseudospiralis* larvae per mouse, respectively.

Samples of blood and spleens were obtained on days: 0 (prior infection), 5, 10, 15, 20, 30, 45 and 60 post infection (p.i.) from all groups, from 5 infected and 3 uninfected mice. The blood was centrifuged at 3000 *g* for 10 minutes and serum samples were stored at -20 °C until the examination.

The Infective Larvae *Trichinella* Sppspp.

The reference isolates of *Trichinella spiralis* (ISS 004), *T. britovi* (ISS 1088) and *T. pseudospiralis* (ISS 013) (obtained and assigned codes from the Trichinella Reference Centre in Rome), maintained by serial passage in ICR mice at the Institute of Parasitology SAS, were used for the infection. Larvae were released by artificial digestion (1 % pepsin, 1% HCl for 4 h at 37 °C; both Sigma-Aldrich, Germany) of tissue following the standard protocol and kept saline solution until inoculation of experimental mice.

Intestinal Worm Burdens

The intestinal phase of infection was investigated on days 5, 10, 15 and 20 p.i. Small intestine was cut into 5-10 cm long pieces, placed into a sieve and incubated in conical pilsner glasses in 37 °C NaCl (0.9 % saline) overnight. After incubation, gut pieces were discarded and the sediment was counted under stereomicroscope at 60 x magnification.

Isolation of Muscle Larvae

The muscle phase of infection was examined on days 20, 30, 45 and 60 p.i. Whole eviscerated carcasses were minced and artificially digested (1 % pepsin, 1 % HCl for 4 h at 37 °C; both Sigma-Aldrich, Germany) according to Kapel and Gamble (2000). Samples were allowed to settle for 20 min before the supernatant was discard and the sediment was poured through a 180 μm sieve into a conical glass and washed with tap water. The sediment was finally transferred to a gridded Petri dished dishes and counted using a stereomicroscope at 40 x magnification. Depending on the density of larvae either a sub-sample or the whole sample was counted.

T and B Lymphocyte Proliferation Assay

The proliferative activity of splenic T and B was detected spectrofotometrically using MTT assay (Dvorožňáková et al. 2008). Briefly, cells (5×10^6 cells /ml RPMI, Sigma-Aldrich, Germany) were incubated with 10 μg/ml of Concanavalin A (Con A) (T cells) or lipopolysaccharide (LPS) (B cells) (Sigma-Aldrich, Germany) at 37°C in 5 % CO_2 and 85 % humidity for 72 hours. 20 μl aliquots of 3,4-dimethylthiazolyl 2,5-diphenyltetrazolium bromide (MTT) (Sigma-Aldrich, Germany) (0.1 % solution) were then added to the cell cultures, incubated for 4 h at 37°C and 5 % CO_2 followed by centrifugation at 800 x g for 5 min. Reaction was terminated with dimethylsulfoxide (Sigma-Aldrich, Germany) (200 μl/cell sample) and read at a 540 nm and 630 nm. The stimulation indices (SI) were calculated according to the formula: SI= $E_{540} - E_{630}$ (stimulated cells) / $E_{540} - E_{630}$ (unstimulated cells).

Number of CD4 and CD8 T Cells

Lymphocytes from the spleen and depleted of erythrocytes were resuspended in PBS (pH 7.2) at a final concentration of 1×10^6 cells /ml. The cellular subpopulations were detected by use of rat anti-mouse CD4+ fluorescein isothiocyanate-conjugated and rat anti-mouse CD8+ phycoerythrin-conjugated monoclonal antibodies (BD Biosciences PharMingen, Belgium) at the concentration of 0.4 μg/10^6 cells at 4°C for 30 min. Cells were then washed three times with PBS containing 0.1% NaN_3 and analysed by the FACScan flow cytometer (Becton Dickinson Biosciences, Germany). All data files were analysed with CellQuest software. The final numbers of both cell populations were calculated as proportion from the total isolated lymphocytes per spleen/mouse.

Concentration of IL-5, IL-10, and IFN-γ *In Vitro* Production

The capture ELISA was employed to determine the concentration of *in vitro* production of cytokines by splenocytes according to the method of Šoltýs and Quinn (1999). Splenocytes for cytokine production were resuspended to 10^7 cells/ml RPMI 1640 medium (Sigma-Aldrich, Germany) and incubated with 10 μg/ml of Con A at 37°C in 5 % CO_2 and 85 % humidity. Supernatants were harvested after 72 hours and stored at −80 °C prior cytokine determinations. Interferon-γ (IFN-γ) and interleukin-5 and interleukin-10 (IL-5, IL-10) were used as marker cytokines for Th1 and Th2 responses, respectively. The pairs of cytokine-specific monoclonal antibodies used were: R4-6A2 and XMG1.2 for IFN-γ; TRFK5 and TRF4 for IL-5; JES5-2A5 and SXC-1 for IL-10 (BD Biosciences PharMingen, Belgium). Results were expressed at pg/ml using murine recombinant IFN-γ, IL-5 and IL-10 (BD

Biosciences PharMingen, Belgium) as standards. The detection limit of the assay for cytokines was 0.04 ng/ml.

Detection of Specific Antibody Production by iELISA

Specific *Trichinella* spp. antibodies in serum were detected by indirect ELISA. Somatic antigens *(T. spiralis, T. britovi, T. pseudospiralis)* diluted at 2 µg/ml carbonate buffer (pH 9.6) were bound to the microtitrate plates (Nunc, Denmark) overnight at 4 °C. After triple washing of wells with phosphate buffered saline (PBS, pH=7.2) with 0.5 % Tween 20 (PBS-T) non-specific bonds were blocked with by 0.5 % skimmed milk PBS after 1 hour incubation at room temperature. After triple washing with PBS-T the serum samples and conjugates were added step by step for 1 hour incubation at 37 °C. Sera were diluted 1:100 in PBS-T. Anti-mouse horseradish peroxidase conjugates (all Sigma-Aldrich, Germany) were diluted: IgM (1:2000), IgG_1 (1:2000), IgG_{2a} (1:500) a IgG_{2b} (1:500). The substrate o-phenylene diamine (Sigma-Aldrich, Germany) at 0,05 mol/l in citrate buffer (pH 4.7) with 0,005 % H_2O_2 was used for a visual reaction. The reaction was stopped by 1M H_2SO_4 after 20 minutes incubation at room temperature in the dark. Plates were measured for the optical density at 490 nm (Revelation Quicklink, Opsys MR, Dynex Technologies, USA).

Statistical Evaluation

Statistical differences were assessed using one-way ANOVA, followed by post hoc Tukey's test (a value of P<0.05 was considered significant), which allowed comparison between each two groups at each time point. The analyse were performed using the Statistica 6.O (Stat Soft, Tulsa, USA) statistical package.

RESULTS

Parasite Burden – Numbers of Adults and Muscle Larvae (Tab. 1)

Mice infected with the low dose of larvae absolutely eliminated parasite adults from the small intestine till day 20 p.i. in both encapsulating species. On day 15 p.i. the occurrence of *T. spiralis* adults in the intestine was sporadic (0.13±0.35) in contrast to *T. britovi*, that were found in higher numbers of worms (2.43±2.51). *T. pseudospiralis* adults were isolated from the small intestine also on day 20 p.i. (0.62±0.74).

Numbers of muscle larvae obtained after *T. spiralis* infection were the highest of all infections, what is related to the high reproductive capacity of this species. The maximum numbers of *T. spiralis* larvae were found on day 45 p.i. (3060 larvae/mice), numbers of *T. britovi* and *T. pseudospiralis* peaked on day 30 p.i. (847.2 a 645.5/mice, respectively). A common biological characteristic used to evaluate *Trichinella* species infectivity is the index of reproductive capacity (RCI) – ratio of the total number of larvae recovered from an animal to the number of larvae administered to it. The first is the product of 4 components:

the number of females established; their fecundity; their reproductive life span; the survival of muscle larvae. At the muscle phase of *T. spiralis* infection RCI reached the values 20.4 – 306.0, that were significantly above the values found in *T. britovi* (1.1 – 84.7) or *T. pseudospiralis* (0.6 – 64.6) infections (Tab. 1).

Table 1. Parasite burden of mice infected with 10 larvae of *Trichinella spiralis*, *T. britovi* and *T. pseudospiralis*. Index of reproductive capacity (RCI) – ratio of the total number of larvae recovered from an animal to the number of larvae administered to it

Intestinal phase - numbers of adults					
Day p.i.	*T.spiralis* mean±S.D.		*T.britovi* mean±S.D.		*T.pseudospiralis* mean±S.D.
5	6.50±2.07		6.83±1.94		3.67±1.51
10	4.00±1.67		5.50±2.07		5.00±2.28
15	0.13±0.35		2.43±2.51		0.83±0.41
20	0		0		0.63±0.74
Muscle phase - numbers of muscle larvae					
Day p.i.	*T.spiralis* mean±S.D.	RCI	*T.britovi* mean±S.D.	RCI	*T.pseudospiralis* mean±S.D.	RCI
20	204.2±84.5	20.4	11.2±4.3	1.1	5.5±8.5	0.6
30	2380.0±1214.8	238.0	847.2±609.6	84.7	645.5±269.1	64.5
45	3060.0±1859.1	306.0	368.6±279.0	36.8	546.2±465.2	54.6
60	1800.0±1182.2	180.0	642.6±239.8	64.3	536.4±295.5	55.4

S.D. statistical deviation.
RCI reproductive capacity index.

Proliferative Response of T Lymphocytes to ConA (Fig. 1)

At the intestinal phase of *T. spiralis* aj and *T. britovi* infections, the stimulated proliferation of T lymphocytes was observed from day 10 to 15 p.i. Values of stimulation indices in *T. spiralis* infection remained increased till the end of the experiment (day 60 p.i.; P<0.05), but *T. britovi* only till day 45. p.i. *T. pseudospiralis* infection induced a rise during the intestinal phase only on day 10 p.i., but the proliferative activity of T cells was increased during newborn larvae migration, from day 30 p.i. (P<0.05) till the end of the experiment.

Proliferative Response of B Lymphocytes to LPS (Fig. 2)

The proliferative activity of B lymphocytes was increased after *T. spiralis* and *T. britovi* infections on day 5 p.i. Then *T. britovi* infection induced only one growth of values on day 15 p.i., but *T. spiralis* infected mice showed a rise in stimulation indices from day 45 p.i. (P<0.05), with the maximum on day 60 p.i. (P<0.01). *T. pseudospiralis* infection significantly stimulated the proliferation of B cells early on day 10 p.i. (P<0.05) and once more from day 30 to 45 p.i. (P<0.05).

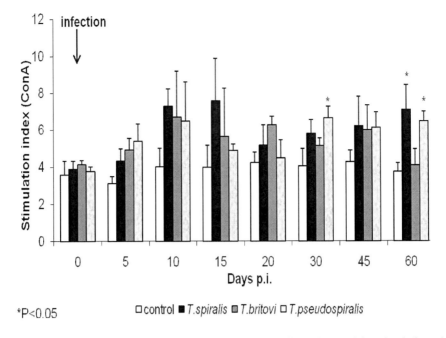

Figure 1. Proliferative response of T lymphocytes to concanavalin A (Con A) in mice infected with 10 larvae of *Trichinella spiralis*, *T. britovi* and *T. pseudospiralis*. *(P<0.05) statistically significant from control uninfected mice.

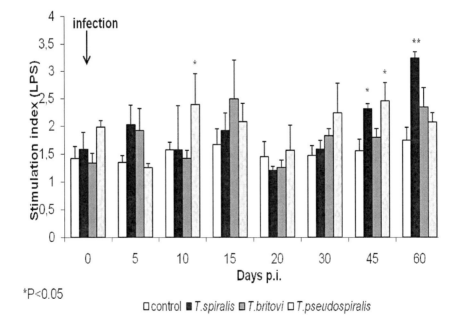

Figure 2. Proliferative response of B lymphocytes to lipopolysaccharide (LPS) in mice infected with 10 larvae of *Trichinella spiralis*, *T. britovi* and *T. pseudospiralis*. *(P<0.05); **P<0.01) statistically significant from control uninfected mice.

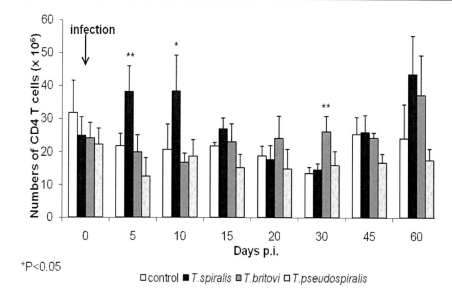

Figure 3. Numbers of splenic CD4 T cells in mice infected with 10 larvae of *Trichinella spiralis, T. britovi* and *T. pseudospiralis*. *(P<0.05); **P<0.01) statistically significant from control uninfected mice.

Numbers of Splenic CD4+ and CD8+ T Lymphocytes (Figs. 3, 4)

In comparison to control, there were not found differences in CD4 and CD8 T cell subpopulations in mice infected with *T. pseudospiralis*. Also *T. britovi* did not markedly affect these cells, with the exception of day 30 p.i., when both CD4 and CD8 T subpopulations almost doubled their numbers. On the contrary, *T. spiralis* significantly increased the occurrence of CD4 and CD8 T cells in the spleen from day 5 to 10 p.i. (P<0.05; P<0.01) and the numbers increased again on day 60 p.i.

In Vitro Production of IL-5, IL-10 and IFN-γ Cytokines (Figs. 5, 6, 7)

The infections of mice with encapsulating *T. spiralis* and *T. britovi* significantly stimulated production of Th2 cytokine IL-5 as early as day 5 or 10 p.i. (P<0.01; P<0.05), respectively. A great increase in the IL-5 production was recorded in both infections during newborn larvae migration and at the muscle phase, from day 15 p.i. with peaks at days 30 and 60 p.i. (P<0.01). The non-encapsulating *T. pseudospiralis* did not influence the IL-5 production at the intestinal phase. At the muscle phase, the IL-5 production was increased, however, reaching the half of IL-5 values found in *T. spiralis* and *T. britovi* infections. (Fig. 5)

The production of anti-inflammatory cytokine IL-10 was inhibited in all infections at the intestinal phase from day 5 to 10 p.i. (P<0.05). *T. britovi* infection stimulated IL-10 on day 15 p.i. Encapsulating trichinellas did not affect the IL-10 synthesis at the muscle phase. On the contrary, *T. pseudospiralis* increased the IL-10 production with the significant stimulation from day 20 till the end of the infection (P<0.05; P<0.01). (Fig. 6).

Pro-inflammatory Th1 response was suppressed after the infections with encapsulating trichinellas, the IFN-γ production was decreased after *T. spiralis* a *T. britovi* infections from day 5 to 10 p.i. The IFN-γ generation started to increase with newborn larvae migration from day 15 p.i. *T. pseudospiralis* infection induced a high IFN-γ production as early as day 10 p.i., which persisted at high values (P<0.01; P<0.05) also at the muscle phase, with the maximum on day 60 p.i. (P<0.05). (Fig. 7).

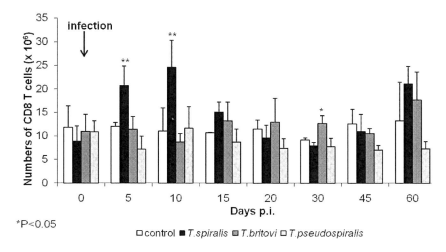

Figure 4. Numbers of splenic CD8 T cells in mice infected with 10 larvae of *Trichinella spiralis*, *T. britovi* and *T. pseudospiralis*. *(P<0.05); **P<0.01) statistically significant from control uninfected mice.

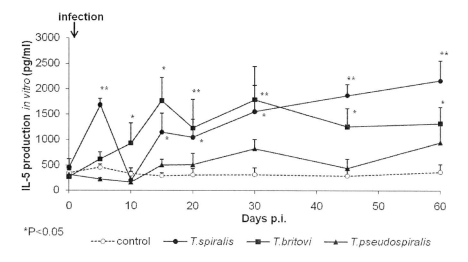

Figure 5. IL-5 production *in vitro* by splenocytes of mice infected with 10 larvae of *Trichinella spiralis*, *T. britovi* and *T. pseudospiralis*. *(P<0.05); **P<0.01) statistically significant from control uninfected mice.

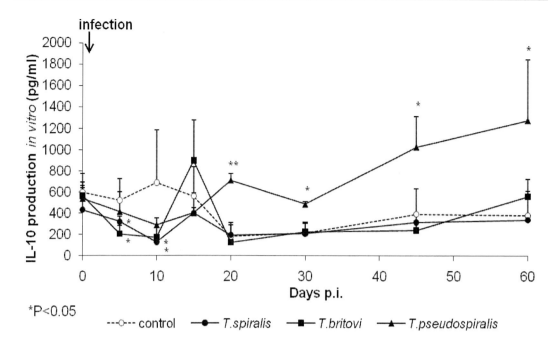

Figure 6. IL-10 production *in vitro* by splenocytes of mice infected with 10 larvae of *Trichinella spiralis*, *T. britovi* and *T. pseudospiralis*. *(P<0.05); **P<0.01) statistically significant from control uninfected mice.

Dynamics of Specific Immunoglobulines IgM, IgG$_1$, IgG$_{2a}$, IgG$_{2b}$ (Figs. 8, 9, 10, 11)

The low infective dose of 10 larvae of *Trichinella spiralis* and *T. britovi* did not induce an increase in specific IgM antibody response, no seroconversion was observed in comparison with the control. However, *T. pseudospiralis* infection evoked the increase IgM immunoglobuline level in murine serum from days 5 to 20 p.i., i.e. during the intestinal phase of the infection. (Fig. 8).

The low infective dose of *Trichinella spiralis* stimulated the production of specific IgG$_1$ antibodies from day 20 p.i., with a sharp increase from day 45 p.i., but *T. britovi* infection induced the generation of specific IgG$_1$ antibodies only on day 60 p.i. *T. pseudospiralis* infection did not affect the IgG$_1$ production. (Fig. 9).

The production of IgG$_{2a}$ and IgG$_{2b}$ was again earlier and more expressive after *T. spiralis* infection from day 45 p.i., on the contrary to *T. britovi*, where the antibody concentration in serum were increased only on day 60 p.i. The low infective dose of *T. pseudospiralis* did not influence IgG$_{2a}$ antibody response, but IgG$_{2b}$ immunoglobuline level started increasing from day 45 p.i., although with a lower intensity as in infections with encapsulating species. (Figs. 10, 11).

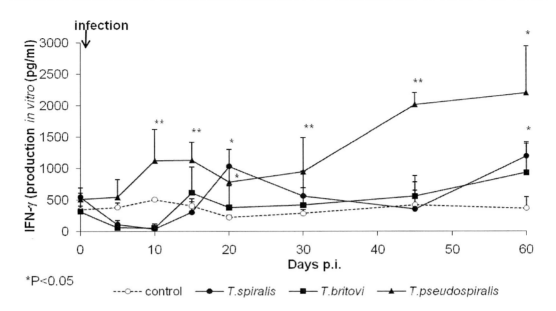

Figure 7. IFN-gamma production *in vitro* by splenocytes of mice infected with 10 larvae of *Trichinella spiralis*, *T. britovi* and *T. pseudospiralis*. *(P<0.05); **P<0.01) statistically significant from control uninfected mice.

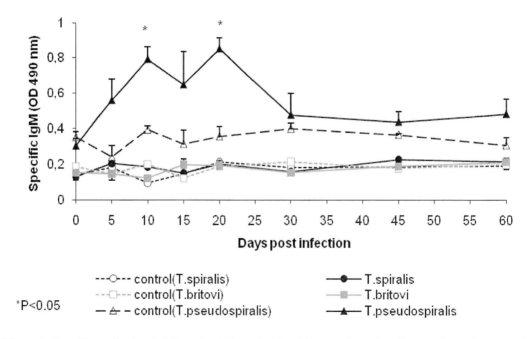

Figure 8. Specific antibodies IgM in mice infected with 10 larvae of *Trichinella spiralis*, *T. britovi* and *T. pseudospiralis*. *(P<0.05) statistically significant from control uninfected mice.

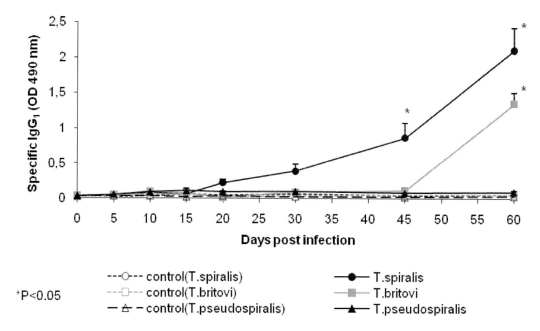

Figure 9. Specific antibodies IgG$_1$ in mice infected with 10 larvae of *Trichinella spiralis*, *T. britovi* and *T. pseudospiralis*. *(P<0.05) statistically significant from control uninfected mice.

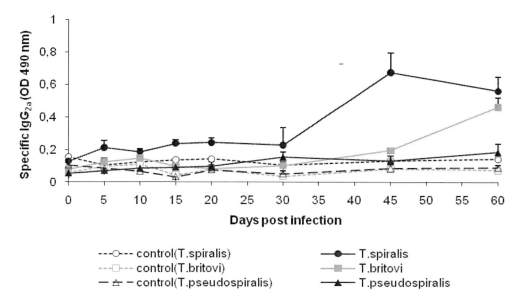

Figure 10. Specific antibodies IgG$_{2a}$ in mice infected with 10 larvae of *Trichinella spiralis*, *T. britovi* and *T. pseudospiralis*.

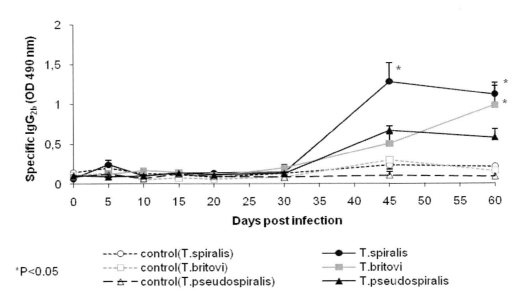

Figure 11. Specific antibodies IgG$_{2b}$ in mice infected with 10 larvae of *Trichinella spiralis*, *T. britovi* and *T. pseudospiralis*. *(P<0.05) statistically significant from control uninfected mice.

DISCUSSION

Trichinella spp., the causative agent of trichinellosis, occupies two distinct intracellular niches within its host, the intestinal epithelium and the skeletal muscle (Despommier 1998), where the interaction between the parasite and the host muscle is unusual for the biology of nematodes. Encapsulating and non-encapsulating trichinellas differ in the nurse cell complex, that is or not surrounded by a collagen capsule (Sacchi et al. 2001). There are several differences in biological and immunological characteristics between encapsulating and non-encapsulating *Trichinella* species, most notably in host inflammatory response and modulation of host phenotype (Li and & Ko 2001; Wu et al. 2001, ; Boonmars et al. 2005), which suggest that the parasite may influence the host immune response in different ways.

This study was focused on differences in infectivity and immunogenicity of three *Trichinella* species (*T. spiralis, T. britovi,* and *T. pseudospiralis*). In experiments, the low infective dose of trichinella larvae was used to simulate natural conditions for the infection of small mammals, the important reservoir hosts of trichinellosis (Antolová et al. 2006; Reiterová et al. 2009). There was studied how the parasite variation influenced the development of cellular responses necessary for host's resistance against the infection. In order to compare infectivity of *T. spiralis, T. britovi* and *T. pseudospiralis* in this study the parasite burden was found. The host expulsion of adults from the small intestine was finished by day 20 p.i. for encapsulting species and day 30 p.i. for *T. pseudospiralis*. A similar trend of worm persistence, a shorter lasting (11 days) for *T. spiralis* and a longer lasting (14 days) for *T. pseudospiralis,* was observed by Furze and Selkirk (2005), who used 500 larvae for infection of mice. Development from the immature larva to the infective L1 is dependent on infection of skeletal muscle; larvae that penetrate cells are developmentally arrested, implying the parasite is able to recognise the host cell environment (Despomier 1998). At the muscle phase of infections *T. spiralis* reached the highest numbers of larvae, what is related to the high reproductive capacity of this species, numbers of *T. britovi* and *T. pseudospiralis* larvae were lower by one arithmetic series. The highest number of *T. spiralis* larvae ultimately establishing in muscles represents the highest antigenic challenge for its host.

In our experimental work the increased proliferative activity of splenic T lymphocytes was recorded at the intestinal phase of encapsulating species infections (*T. spiralis* a *T. britovi*) from day 10 to 15 p.i. *T. pseudospiralis* infection (non-encapsulating species) stimulated the proliferative response of T cells only at day 10 p.i. even though adults were obtained from the small intestine even at day 20 p.i. Maybe a small activated function of lymphocytes, except for other effector components of immunity (mastocytes, macrophages), allowed a longer persistance of *T. pseudospiralis* in the host intestine. A suppressed inflammation has been attributed to *T. pseudospiralis* by works (Stewart et al. 1985; Furze and & Selkirk 2005). At the muscle phase all three trichinella species induced an increased the T cell proliferation. The highest stimulation was detected in *T. spiralis* infection and remained till the end of the infection, *T. britovi* infection activated the T cell proliferation only till day 45 p.i. *T. pseudospiralis*, despite of low numbers of muscle larvae, increased the T cell proliferative activity from day 30 p.i. to 60 p.i. Probably the low infective dose did not cause significant changes in the B cell proliferation at the intestinal phase as B cell mitosis was activated only on day 5 or 10 p.i. The values of stimulation indices started to increase at the developed muscle phase, from day 45 p.i. , with the highest rise in *T. spiralis* infection. Polyclonal lymphocyte activation of T-cells, but particularly B-cells, is responsible for the high levels of immunoglobulines IgG, IgM, and IgE (Murrel and & Bruschi, 1994). In our experiment the proliferation of splenic B cells was stimulated only at the muscle phase, from day 45 p.i., when also seroconversion for specific antibodies was recorded.

A functional status of T lymphocytes is related to two subpopulations (helper CD4 and cytotoxic CD8 cells). The numbers of these cells did not significantly influence by *T. pseudospiralis* infection. *T. britovi* increased the CD4 and CD8 cell numbers only at the advanced muscle phase, with the maximum on day 30 p.i., when the highest numbers of muscle larvae were isolated. Only *T. spiralis* significantly increased the proportion of CD4 and CD8 cells in the spleen at both the intestinal and muscle phases. During the host immune intestinal response, CD4 T cells play a key role in trapping and removal of intestinal worms from the gut, they mediate mucosal changes including intestinal goblet cell hyperplasia with high mucin secretion (Ishikawa et al. 1997). CD4 T cells are also related to a reduction of naive cells and an increased generation of memory cells (Morales et al. 2002). CD8 T cells play no significant role in worm expulsion (Vallance et al. 1999), although their proportion increased at the intestinal phase of *T. spiralis* infection.

Cytokines synthetized primarily by T helper cells play a dominant role in orchestrating both anti-parasite responses and pathology in many helminth infections including trichinellosis (Finkelman et al. 1997). In order to examine a relation between Th1 versus Th2 cytokine profiles, levels of "marker" cytokines were monitored. Furze and Selkirk (2005), who infected mice with 500 larvae, recorded a mixed Th1/Th2 cytokine expression from spleen cells with no qualitative differences between *T. spiralis* and *T. pseudospiralis* infections. In our work a significant stimulation of *in vitro* production of Th2 cytokine – IL-5 was found at the intestinal *T. spiralis* and *T. britovi* infections. The continual IL-5 growth was recorded in *T. britovi* infection also during newborn larvae migration and at the muscle phase with the maximum on day 30 p.i., when the highest numbers of muscle larvae were found. The IL-5 after *T. spiralis* infection reached the highest values among all three infections with the maximum on day 60 p.i. This species is characterized by the highest reproductive capacity with the highest number of larvae. The enhancement of IL-5 reflected a predominant Th2 immune response in low *Trichinella* spp. infection. Non-encapsulating *T. pseudospiralis* increased IL-5 synthesis only at the muscle phase, however at much lower concentration in comparison to encapsulating trichinellas. Cytokine IL-5 indicates the severity of inflammation, because it directly induces degranulation of eosinophils (Horie et al. 1996).

The production of cytokine IL-10 was moderately inhibited at the intestinal phase of all three *Trichinella* spp. infections, from day 5 to 10 p.i. On the contrary to encapsulating trichinellas, *T. pseudospiralis* stimulated production of IL-10, the cytokine with anti-inflammatory effect, at the muscle infection. Positive suppresion of inflammation induced by *T. pseudospiralis* was described in multiple sclerosis in rats (Boles et al. 2000).

Pro-inflammatory Th1 response was inhibited at the intestinal phase of infections with encapsulating trichinellas. *T. pseudospiralis* stimulated IFN-γ synthesis from day 10 p.i. and induced the highest cytokine concentration in comparison to encapsulating trichinellas. Migration of *T. spiralis* newborn larvae increased the IFN-γ production similarly to infections with a high infective dose of *T. spiralis* larvae by other authors who recorded higher levels of serum IFN-γ between days 7 and 20 p.i. (Frydas et al. 1996, ; Dzik et al. 2002). IFN-γ is crucially involved in protection against newborn larvae, but does not affect the expulsion of adult worms (Helmby and & Grencis, 2003). Its increased production confirms its participation in immune response at the muscle phase. IFN-γ has an antagonistic cytokine IL-10. IFN-γ inhibits macrophage secretion of IL-10 (Mosmann 1994). The balance between IL-10 and IFN-γ determines the development of immunity against the life stages of the parasite (Helmby and & Grencis 2003). Beiting et al. (2004) revealed a role for IL-10 in limiting inflammatory responses during the early stages of muscle infection by *T. spiralis*, but the chronic inflammation is independent on IL-10 and is accompanied by a shift to a Th2 response following completion of parasite development in the muscle. In our work, the high IL-10 production after *T. pseudospiralis* infection persisted also in the late muscle infection.

The presence of immune complexes in the vascular system is connected related with to the symptoms of trichinellosis at the beginning of the infection (Feldmeier et al., 1987). Later, when the parasite has been settled in muscles, neither *Trichinella* antigens nor immune complexes are detectable in blood, but there are detectable specific antibodies. This phenomenon is caused by excretory-secretory antigens from larvae settled in muscles (Pritchard, 1985; Li et al., 1999). Humoral immune response is important in the host defence against migrating newborn larvae. The absence of antigen on the cuticular surface of adult worms is in a sharp contrast to the findings described in muscle larvae, whose surface antigenicity is identical to that of stichocyte α-granules (Appleton et al., 1991). Therefore, no direct immune attack is likely to be exerted against adults. It has been suggested that the effects of serum antibodies on worms may be indirect and that worm expulsion is the result of inflammatory processes in the intestines evoked by the infection (Wakelin & Wilson, 1979).

An increased specific IgM antibody response, which is typical for acute infection, was not found in our experiment after the low infective dose of 10 larvae of *T. spiralis* and *T. britovi*. In comparison with control, neither seroconversion was detected. On the contrary, Reiterová et al. (2009) infected mice with 50 larvae of *T. spiralis* and recorded the seroconversion on day 30 p.i., but IgM antibody production overdrew a cut-off only a little. The low number of antigens might be bound into immune complexes. Contrary, *T. pseudospiralis* induced a higher level of specific IgM immunoglobulines during the intestinal phase till day 30 p.i., what reflects the acute stage of the infection.

An elevation of IgG_1 accompanies the muscle phase of infection (Doligalska, 2000) and newborn larve are more sensitive *in vitro* to IgG_1 in antibody-dependent cellular cytotoxicity (Moskwa, 1999). The low infective dose of 10 larvae of *T. spiralis* in our study stimulated IgG_1 antibody production from day 20 p.i., with a significant elevation from day 45 p.i., but *T. britovi* infection induced the generation of IgG_1 antibodies not until day 60 p.i. It correlates to IgG antibody response in outbred ICR mice after *T. spiralis* infection with low dose of 5 larvae (Reiterová et al., 2009). *T. pseudospiralis* infection did not show a positive specific IgG_1 antibody response. The production of specific IgG_{2a} and IgG_{2b} was again more expressive and earlier after *T. spiralis* infection from day 45 p.i. in contrast to *T. britovi* infection, where these antibodies increased their serum levels not until day 60 p.i. Only isotype IgG_{2b} was detected in *T. pseudospiralis* infection on days 45 and 60 p.i., however at very low values in comparison to encapsulating *Trichinella* species.

Study by Furze and Selkirk (2005) compared antibody response in mice after *T. spiralis* a *T. pseudospiralis* infection with 500 larvae, whereby all classes of parasite-specific antibody were present in serum, but there were differences in the timing. Infection with *T. spiralis* notably induced greater amounts of IgM, IgG_1, IgG_{2b} and IgG_3 in serum during the muscle phases of infection.

The obtained results suggest species-specific differences in immunogenicity of *T. spiralis*, *T. britovi* and *T. pseudospiralis*. Although *T. spiralis* and *T. britovi* showed the same trend, they differed in an intensity of the host immune response. The most pathogenic *T. spiralis* elicited stronger T-cellular response at the intestinal phase of the infection, which resulted in a faster elimination of adults from the host gut. However, a higher reproductive capacity of *T. spiralis* caused a more massive migration of newborn larvae in comparison to *T. britovi* and *T. pseudospiralis*. The lowest immunogenicity was shown by *T. britovi*, which corresponds to its low pathogenicity. However, despite the low infective dose, the host immune system reacted to such a weak antigenic stimul. The strongest cellular response related to increased proliferation of T cells and their differenciation into helper CD4 T lymphocytes was found after *T. spiralis* infection. Both encapsulating trichinellas induced the development of Th2 cytokine response by the intestinal phase and Th2 dominant role was confirmed also at the advanced muscle phase. The biggest deviations in the development of host immune response were shown by non-encapsulating species *T. pseudospiralis*. Decreased numbers of differenciated CD4 and CD8 T cells suggest a suppression of the cellular response and explain an absence of Th2 response at the intestinal phase of the infection, when *T. pseudospiralis* adults survived in the host gut for a longer time in comparison to *T. spiralis* and *T. britovi*. The immune response to *T. pseudospiralis* was directed more to the Th1 type of immune reactions (increased IFN-γ production) at the muscle phase of the infection. It might be related to a longterm diffuse myopathy in *T. pseudospiralis* infection, where Boonmars et al. (2005) observed delayed desintegration of muscle cells in comparison to *T. spiralis* infection. However, in our experiment, there was recorded a high production of both antagonistic cytokines IL-10 and IFN-γ and such cytokine unprofiled environment may provide a longterm survival of *T. pseudospiralis* larva.

The low infective dose of *T. spiralis*, *T. britovi*, and *T. pseudospiralis* in our study induced a late seroconversion in infected mice. Interspecies differences were found in immunogenicity

The low infective dose of *T. spiralis*, *T. britovi*, and *T. pseudospiralis* in our study induced a late seroconversion in infected mice. Interspecies differences were found in immunogenicity of *T. spiralis* and *T. britovi* , which showed the similar trend, but varied in the intensity of the host antibody response. The species *T. spiralis* appeared to be more immunogenic and evoked more intensive and earlier specific antibody response of the host from day 45 p.i., when the antigen material had been markedly accumulated from the newborn and muscle larvae. As the reproductive capacity of *T. britovi* and *T. pseudospiralis* is lower in contrast to *T. spiralis*, the start of specific antibody response in *T. britovi* and *T. pseudospiralis* infections was recorded from day 60 p.i. The immune response to *T. pseudospiralis* infection suggested the biggest deviations. *T. pseudospiralis* adults persisted in the host small intestine for a longer time, untill day 20 p.i. Only mice infected with *T. pseudospiralis* reacted by low IgM antibody production during the intestinal phase, which was not observed in infections with encapsulating *Trichinella* species.

All species *T. spiralis*, *T. britovi* and *T. pseudospiralis* are intracellular parasites of muscle cells, essentially with the same life cycle, which differ in only the fact that complex "nurse cell" is or is not surrounded with a collagen capsule. The differences found in the host immune reactions to different *Trichinella* species in our study suggest that the immune response variations can be caused not only with distinctions of muscle larval L1 excretory-secretory antigens (Robinson et al. 2007; Milcheva et al. 2009), but also with parasite stages L2-4 and adults. Identification and characterization of species-specific proteins of development stages play an important role for understanding of mechanisms involved in the parasite-host interaction, which provide a longterm surviving for the parasite in the host organism.

ACKNOWLEDGMENTS

This study was supported by the Slovak VEGA agency, grants No. 2/0071/08 and No. 2/0093/11. The experimental protocols complied with the current Slovak ethic law.

REFERENCES

Andrade MA, Siles-Lucas M, Lopez-Aban J, Nogal-Ruiz JJ, Perez-Arellano JL, Martinez-Fernandez AR, Muro A (2007) *Trichinella*: Differing effects of antigens from encapsulated and non-encapsulated species on *in vitro* nitric oxide production. *Vet Parasitol* 143:86-90

Antolová D, Reiterová K, Dubinský P (2006) The role of wild boars (*Sus scrofa*) in circulation of trichinellosis, toxocarosis and ascariosis in the Slovak Republic. *Helminthologia* 43:92-97

Appleton JA, Bell RG, Homan W, VanKnapen F (1991) Consensus on *Trichinella spiralis* antigens and antibodies. *Parasitol Today* 7:190-192

Appleton JA, Usack L (1993) Identification of potential antigenic targets for rapid expulsion of *Trichinella spiralis*. *Mol Biochem Parasit* 58:53-62

Beiting DP, Bliss SK, Schlafer DH, Roberts VL, Appleton JA (2004) Interleukin-10 limits local and body cavity inflammation during infection with muscle-stage *Trichinella spiralis*. *Infect Immun* 72:3129-3137

Bolas-Fernández F. (2003) Biological variation in *Trichinella* species and genotypes. *J. Helminthol* 77:111-118

Boles LH, Montgomery JM, Morris J, Mann MA, Stewart GL (2000) Suppression of multiple sclerosis in the rat during infection with *Trichinella pseudospiralis*. *J Parasitol* 86:841-844

Boonmars T, Wu Z, Nagano I, Takahashi Y (2005) *Trichinella pseudospiralis* infection is characterized by more continous and diffuse myopathy than *T. spiralis* infection. *Parasitol Res* 97:13-20

Bruschi F, Tassi C, Pozio E (1992) Parasite-specific antibody response in *Trichinella sp.* 3 human infection: a one year follow-up. *Am J Trop Med Hyg* 43:186 – 193

Campbell WC (1983) Epidemiology I. Modes of transmission. In CampBell WC (Ed.) *Trichinella and Trichinosis*. Plenum Press, New York (U.S.A.), pp. 425 – 444

Candolfi E, Franche P, Liance M, Houin R, Vien T (1989) Detection of circulating antigen in trichinellosis by immunology. Comparative results in mice, rats and humans. Trichinellosis. *Proceedings of the 7th International Conference on Trichinellosis,* pp. 184 – 201

Cui J, Wang ZQ, Han HM (2006) Congenital transmission of *Trichinella spiralis* in experimentally infected mice. *Helminthologia* 43:7-10

Despommier DD (1998) How does *Trichinella spiralis* make itself at home? *Parasitol Today* 14:318-323

Doligalska M (2000) Immune response to *Trichinella spiralis* larvae after treatment with the anti-allergic compound ketotifen. *Parasitol Res* 86: 232–238

Dvorožňáková E, Porubcová J, Šnábel V, Fedoročko P (2008) Immunomodulative effect of liposomized muramyltripeptide phosphatidylethanolamine (L-MTP-PE) on mice with alveolar echinococcosis and treated with albendazole. *Parasitol Res* 103:919-929

Dziemian E, Machnicka B (2000) Influence of *Trichinella spiralis* infective dose on the level of antibodies, circulating antigens and circulating immune complexes in rats. *Helminthologia* 37:59–66

Dzik JM, Golos B, Jagielska E, Kapala A, Walajtys-Rode E (2002) Early response of guinea-pig lungs to *Trichinella spiralis* infection. *Parasite Immunol* 24:369-379

Else KJ, Finkelman FD (1998) Intestinal nematode parasites, cytokines and effector mechanisms. *Int J Parasitol* 28:1145–1158

Feldmeier H, Fischer H, Blaumeiser G (1987) Kinetics of humoral response during the acute and the convalescent phase of human trichinosis. *Zbl Bacter Mikrob Hyg Series A-Med Microb Inf Dis Virol Parasitol* 264:221–234

Finkelman FD, Sheadonohue T, Goldhill J, Sullivan CA, Morris SC, Madden KB, Gause WC, Urban JF (1997) Cytokine regulation of host defense against parasitic gastrointestinal nematodes: Lessons from studies with rodent models. *Ann Rev Immunol* 15:505-533

Frydas S, Karagouni E, Dotsika E, Reale M, Barbacane RC, Vlemmas I, Anogianakis G, Trakatellis A, Conti P (1996) Generation of TNF alpha, IFN gamma, IL-6, IL-4 and IL-

10 in mouse serum from trichinellosis: effect of the anti-inflammatory compound 4-deoxypyridoxine (4-DPD). *Immunol Lett* 49:179-84

Furze RC, Selkirk ME (2005) Comparative dynamics and phenotype of the murine immune response to *Trichinella spiralis* and *Trichinella pseudospiralis*. *Parasite Immunol* 27:181-188

Helmby H, Grencis RK (2003) Contrasting roles for IL-10 in protective immunity to different life cycle stages of intestinal nematode parasites. *Eur J Immunol* 33:2382-2390

Horie S, Gleich GJ, Kita H (1996) Cytokines directly induce degranulation and superoxide production from human eosinophils. *J Allergy Clin Immunol* 98:371-381

Ishikawa N, Wakelin D, Mahida YR (1997) Role of T helper 2 cells in intestinal goblet cell hyperplasia in mice infected with *Trichinella spiralis*. *Gastroenterology* 113:542-549

Kapel CMO, Gamble HR (2000) Infectivity, persistence, and antibody response to domestic and sylvatic *Trichinella spp.* in experimentally infected pigs. *Int J Parasitol* 30:215-221

Khan WI, Vallance BA, Blennerhassett PA, Deng Y, Verdu EF, Matthaei KI, Collins SM (2001) Critical role for signal transducer and activator of transcription factor 6 in mediating intestinal muscle hypercontractility and worm expulsion in *Trichinella spiralis*-infected mice. *Infect Immun* 69:838-844

Kolodziej-Sobocinska M, Dvorožňáková E, Dziemian E (2006) *Trichinella spiralis*: Macrophage activity and antibody response in chronic murine infection. *Exp Parasitol* 112:52–62

Lee KM, Ko RC (2006) Cell-mediated response at the muscle phase of *Trichinella spiralis* infections. *Parasitol Res* 99:70-77

Li CKF, Chung YYY, Ko RC (1999) The dictribution of excretory/secretory antigens during the muscle phase of *Trichinella spiralis* and *T. pseudospiralis* infections. *Parasitol Res* 85:993–998

Li CKF, Ko RC (2001) Inflammatory response during the muscle phase of *Trichinella spiralis* and *T. pseudospiralis* infections. *Parasitol Res* 87:708–714

Machnicka B, Prokopowicz D, Dziemian E, Kołodziej-Sobocińska M. (2001) Detection of *Trichinella spiralis* antigens in urine of men and animals. *Wiadomości Parazytologiczne* 47:217–225

Mahida YR (2003) Host-parasite interactions in rodent nematode infections. *J Helminthol* 77:125-131

Malakauskas A, Kapel CM, Webster P (2001) Infectivity, persistence and serological response of nine *Trichinella* genotypes in rats. *Parasite* 8: S216-S222

Milcheva R, Petkova S, Babál P (2009) Detection of O-glycosylated proteins from different *Trichinella* species muscle larvae total extracts. *Helminthologia* 46:139-144

Milcheva R, Petkova S, Babál P (2009) Detection of O-glycosylated proteins from different *Trichinella* species muscle larvae total extracts. *Helminthologia* 46:139–144

Miterpáková M, Hurníková Z, Antolová D, Dubinský P (2009) Endoparasites of red fox (*Vulpes vulpes*) in the Slovak Republic with the emphasis on zoonotic species *Echinococcus multilocularis* and *Trichinella* spp. *Helminthologia* 46:73-79

Morales MA, Mele R, Sanchez M, Sacchini D, De Giacomo M, Pozio E (2002) Increased CD8(+)-T-cell expression and a type 2 cytokine pattern during the muscular phase of *Trichinella* infection in humans. *Infect Immun* 70:233-239

Morgan UM (2000) Detection and characterisation of parasites causing emerging zoonoses. *Int J Parasitol* 30:1407-1421

Moskwa B (1999) *Trichinella spiralis*: in vitro cytotoxicity of peritoneal cells against synchronous newborn larvae of different age. *Parasitol Res* 85:59–63

Mosmann TR (1994) Properties and functions of interleukin-10. *Adv Immunol* 56:1-26

Murrell KD, Bruschi F (1994) Clinical trichinellosis. *Prog Clin Parasitol* 4:117-150

Paraličová Z, Kinčeková J, Schréter I, Jarčuška P, Dubinský P Jr, Porubčin Š, Pavlinová J, Kristian P (2009) Outbreak of trichinellosis in eastern Slovakia. *Helminthologia* 46:209-213

Pritchard DI (1985): Antigen production by encysted muscle larvae of *Trichinella spiralis*. *J Helminthol* 59:71–77

Reiterová K, Antolová D, Hurníková Z (2009) Humoral immune response of mice infected with low doses of *Trichinella spiralis* muscle larvae. *Vet Parasitol* 159:232-235

Robinson MW, Grieg R, Beattie KA, Lamont DJ, Connolly B (2007) Comparative analysis of the excretory-secretory proteome of the muscle larva of *Trichinella pseudospiralis* and *Trichinella spiralis*. *Int J Parasitol* 37:139-148

Ros-Moreno RM, de Armas-Serra C, Gimenez-Pardo C, Rodriguez-Caabeiro F (2002) Comparison of cholinestase activities in the excretion-secretion products of *Trichinella pseudospiralis* and *Trichinella spiralis* muscle larvae. *Parasite* 9:153-159

Sacchi L, Corona S, Gadjadhar AA, Pozio E (2001) Ultrastructural characteristics of nurse cell-larva complex of four species of *Trichinella* in several hosts. *Parasite* 8: S54-S58

Šnábel V, Malakauskas A, Dubinský P, Kapel CMO (2001) Estimating the genetic divergence and identification of three *Trichinella* species by isoenzyme analysis. *Parasite* 8:S30-S33

Šoltýs J, Quinn MT (1999) Modulation of endotoxin- and enterotoxin-induced cytokine release by in vivo treatment with beta-(1,6)-branched beta-(1,3)-glucan. *Infect Immun* 67:244-252

Stewart GL, Wood B, Boley RB (1985) Modulation of host response by *Trichinella pseudospiralis*. *Parasite Immunol* 7:223-233

Turčeková Ľ, Borošková Z, Tomašovičová O, Reiterová K, Kinčeková J (1997) Immunochemical analysis of larval antigens of *Trichinella spiralis* and *T. pseudospiralis*. *Helminthologia* 34:241-243

Vallance BA, Galeazzi F, Collins SM, Snider DP (1999) CD4 T cells and major histocompatibility complex class II expression influence worm expulsion and increased intestinal muscle contraction during *Trichinella spiralis* infection. *Infect Immun* 67:6090-6097

Wakelin D, Farias SE, Bradley JE (2002) Variation and immunity to intestinal worms. *Parasitology* 125:S39-S50

Wakelin D, Wilson MM (1979) *Trichinella spiralis*: immunity and inflammation in the expulsion of transplanted adult worms from mice. *Exp Parasitol* 48:305–312

Wang CH (1997) Study of biological properties of *Trichinella spiralis* newborn larvae and the antiparasitic mucosal immunity of the host. *Frontiers in Bioscience* 2:317-330

Wu Z, Matsuo A, Nakada T, Nagano I, Takahashi Y (2001) Different response of satellite cells in the kinetics of myogenic regulatory factors and ultrastructural pathology after *Trichinella spiralis* and *T. pseudospiralis* infection. *Parasitology* 123:85-94

In: Larvae: Morphology, Biology and Life Cycle
Editors: Kia Pourali and Vafa Niroomand Raad

ISBN: 978-1-61942-662-7
© 2012 Nova Science Publishers, Inc.

Chapter 6

ECONOMICALLY IMPORTANT MARINE LARVAE: AQUACULTURE PRODUCTION AND LARVAL REARING

Julian Gamboa-Delgado[*]

Mariculture Program, Ecology Department, Faculty of Biological Sciences,
Universidad Autónoma de Nuevo León, Cd. Universitaria. Apdo.
San Nicolás de los Garza, Nuevo León, México

ABSTRACT

The production of animals through aquaculture practices has narrowed down the dependence on fisheries-derived products. Aquaculture production yields are increasing but the majority of farmed aquatic animals are still represented by freshwater fish such as carp and tilapia. On the other hand, the production of marine organisms is dominated by several species of fish, crustaceans and molluscs; which have been successfully breed in captivity through their whole life cycles. These advances have been in part supported by applied research that has led to development of well-established rearing practices. The continued growth of the aquaculture industry requires high numbers of good quality postlarvae and juveniles produced in commercial hatcheries. Very often, the larval rearing of marine species represents the main bottleneck of the whole aquaculture production process and this is highlighted by the present situation with the production of marine fish. Although hundreds of species of marine fish are viable for farming, the current production accounts to only 3% of the global production of farmed aquatic animals. This figure is in part explained by the current lack of knowledge on the digestive physiology and nutrient requirements for most marine fish larvae. While mollusc larvae entirely depend on phytoplankton, most economically important species of marine crustaceans and fish are strongly dependent on zooplankton during the larval stage. The intrinsic difficulties in producing large amounts of specific live feeds to support the larval rearing of many marine species represents one of the main problems that marine farmers have to cope with. Therefore, intensive research on the ontogeny and physiology of the larval digestive tract is continuously conducted in order to have a better understanding of the larval digestive processes. New findings on the biology and physiology of marine larvae rapidly impacts technical production aspects such as the shape and size of the

---

[*] e-mail: julian.gamboad@uanl.mx julian.gamboad@uanl.edu.mx, Tel/Fax +52 (81) 8352-6380.

rearing vessels, the establishment of optimal larval rearing conditions and the larval feeding protocols used to supply live and inert feeds. New research findings assist nutritionists to formulate diets that can successfully replace live preys needed during the critical larval rearing stages.

INTRODUCTION

The farming of aquatic plants and animals is collectively known as aquaculture and at the present time represents one of the most dynamic industries with an estimated global growth rate of 8.3% between 1970 and 2008, yielding a bulk production of 52 million metric tons of products in the latter year (FAO, 2010). The production of marine animals or mariculture has also been steadily increasing over the last two decades. It is evident that if the finfish and shellfish aquaculture sectors are to sustain their current growth rate, then the supply of marine larvae will also have to grow at similar rates so as to meet demand. Although aquaculture production has narrowed down the dependence on fisheries-derived products, the majority of farmed aquatic animals are still represented by freshwater fish such as carp and tilapia. On the other hand, the production of marine organisms is dominated by several species of molluscs and crustaceans that have been successfully farmed. The development of well-established farming techniques has been supported by new findings from applied research. However, one of the big challenges the aquaculture industry has yet to overcome is to substantially increase the production of marine organisms. Several species of crustacean, fish and molluscs are now successfully farmed. Reproduction in captivity has been achieved, in this manner closing the life cycles. There are hundreds of marine species viable for aquaculture production and new rearing methodologies are slowly bridging the gaps between research activities and commercial technical feasibility. Although there are several crucial issues that can hamper the aquaculture production process (maturation of broodstock animals in captivity, disease control), the main bottleneck of this activity has been for decades identified as the successful production of sufficient numbers of high quality larvae through the refinement of rearing techniques. Knowledge on basic aspects of the biology of marine larvae is then essential to achieve constant productions. Many research efforts have been devoted to one particular aspect, namely, the study of the nutritional physiology. The latter is considered to be one of the most important issues in marine larvae research (Cahu and Zambonino-Infante, 2001; Koven et al., 2001; Conceição et al., 2010).

MARINE LARVAE REQUIRED BY THE AQUACULTURE INDUSTRY

Only a small percentage of companies devoted to the production of marine organisms have their own maturation and larviculture facilities; therefore, most companies need to satisfy their demand for postlarvae and juveniles by acquiring animals from commercial hatcheries, which in turn operate as a whole, separate industry. Although some marine larval organisms can be collected from the natural environment (*e.g.* oyster and mussel larvae), reliance on commercial hatcheries is critical for some aquaculture industries such as shrimp farming. Shrimp production through aquaculture practices has achieved a high degree of control due to nutritional advancements, selective breeding programs and pathogen management, among others. These advances have been reflected in increasing production yields. For example, in 2008, 46 % of the world production of crustaceans was supplied by the aquaculture industry, most of these were Penaeid shrimp farmed in semi-intensively managed ponds, which in turn are stocked with postlarvae produced in commercial hatcheries. On the other hand, the demand for mollusc larvae, mainly bivalves, is higher than the availability of naturally-occurring larvae that can be seasonally collected (usually by allowing larval setting on artificial substrates placed in the ocean). Bivalve hatcheries may have both, research and commercial-scale larval rearing capabilities and can provide the majority of bivalves (oysters, clams, mussels and scallops) needed by grow out operations of this particular industry (Delmendo, 1990; Utting and Spencer, 1991). Compared to crustacean and mollusc hatcheries, there are significantly less commercial hatcheries producing marine fish larvae. Some representative species being produced are sea bream, sea bass, and Senegal sole in Europe (FAO, 2010), sand whiting, snapper and yellow fin bream in USA (Tucker, 1998) and cobia in Asia (Benetti, 2010). Besides offering constant availability and competitive price, hatcheries also frequently certify their animals as specific pathogen free or specific pathogen resistant (Moss and Arce, 2001).

DIGESTIVE PHYSIOLOGY OF MARINE LARVAE PRODUCED IN HATCHERIES

Shortly after hatching, most marine larvae become transitory individuals of the planktonic ecosystem, where they spent at least one planktonic larval stage of their life cycle. Although marine larvae share the same trophic position, there are clear differences in the digestive larval morphology even among individuals belonging to the same taxa. The main function of the digestive system is to degrade macronutrients from dietary elements into absorbable forms in order to supply nutrients to the body tissues; however, the rudimentary digestive tract of marine larvae has limited capacity to assimilate a wide range of dietary items. Common characteristics of the digestive tract of most marine larvae are its tubular shape, presence of rudimentary accessory digestive glands and a yolk reserve. Shortly after yolk (vitellin) reserves are depleted, the exogenous feeding starts and larvae start consuming phytoplankton and/or zooplankton. This shift from endogenous to exogenous feeding is a very critical transition when rearing marine larvae in hatcheries, as feed has to be readily available and in the proper densities for its uptake. Most marine fish larvae hatch from small, free floating eggs, and the larval digestive tract is still developing at the onset of exogenous feeding. The larval digestive system thus consist of a simple digestive tract having limited gland-containing surface; however, marine larvae have a high capacity for digesting live preys, efficiently degrading and absorbing dietary nutrients needed to support very high growth rates. A continuous availability of high concentrations of suitable food is therefore required for the survival of the larvae since the fully developed digestive system is acquired after larval metamorphosis within a period ranging from weeks to months after first feeding depending on species (Ronnestad et al., 2007). Significant progress has been made developing formulated feeds for fish larvae that result in high survival and growth and low incidence of skeletal malformations; however, these positive results still depend on the supply of co-fed live feeds (Yúfera et al., 2005). Knowledge on the digestive physiology of larval stages of fish is gradually improving. However, the main focus has been on studies describing the morphological and histological development of the digestive organs and on the expression and activity of digestive enzymes (Ronnestad et al., 2007).

Larvae of Decapod crustaceans (marine shrimps, freshwater prawns, crabs and lobsters) metamorphose through a number of remarkable anatomical adaptations such as mouthpart structure and functional morphology of the digestive system (Jones et al., 1997). The development of larval gut follows a similar pattern in Decapods, with differentiated mouthparts used for consumption of different diet types, while the anterior stomach chamber remains simple until the late larval stages. These changes are related to shifts in feeding habits from herbivory to carnivory and, in the particular case of Penaeid shrimps, larvae metamorphose from a non-feeding stage (nauplii) to a phytoplankton-feeding stage (zoea) and finally to a carnivorous stage (mysis) before reaching postlarval stage. Digestive structures such as hepatopancreas are formed during ontogeny by the progressive transformation of several larval structures. Some other structures are progressively lost, for example, the midgut tissue in the naupliar stage of Penaeids, which functions in digestion and absorption of yolk (Lovett and Felder, 1990). The herbivorous zoea stages have greater digestive enzyme-secretory area, relative to gut volume, than larger carnivorous larvae. As larvae develop and switch to carnivory, the secretory area declines (Lovett and Felder, 1990; Abubakr and Jones, 1992). The hepatopancreas and gland filter develop towards the final, adult functional morphology during the later larval stages (Le Vay et al., 2001). As larvae are dependent upon enzymatic breakdown of ingested food, the development of secretory tissue in the hepatopancreas (main digestive gland) dictates the type of prey which can be consumed (See review by Jones et al., 1997). The activities of several digestive enzymes of crustacean larvae have been determined (See review by Le Vay et al., 2001). It has been shown that enzyme activities are highly correlated to the type of diet consumed during specific larval stages; hence, larval shrimps successfully adapt to carnivorous feeding at early developmental stages (zoea). Digestive enzymes activities in crustacean larvae have been found to be high, indicating a strong adaptation to herbivory at first feeding and also indicating modulation of enzyme content in response to dietary quality (Le Vay et al., 2001). Trypsin is the dominant digestive enzyme during larval development (Lovett and Felder, 1990; Kamarudin, 1992). In general, trypsin activity in Penaeid larvae feeding on live feed peaks at the metamorphosis from zoea to mysis stage, thereafter declines towards the first postlarval stage (McDonald et al., 1989; Jones et al., 1993). An adaptive increase in digestive enzyme activity in response to feeding on artificial diets and algal diets of low protein content has been observed, particularly in the second larval stages (mysis) (Jones et al., 1993). Adaptation to carnivorous or omnivorous feeding at early development stages results in lower digestive enzyme levels (Le Vay et al., 2001) but this event it might also be a response to anatomical changes in the digestive tract (Lovett and Felder, 1990b).

Bivalve molluscs are filter-feeders (they take in seawater from the surrounding environment and "filter" out particles present in the water). The larvae feed on phytoplankton as soon as (or shortly before) vitellin reserves are depleted. After the larval stage they are able to continue feeding solely on microalgae or suspended material throughout their life cycle (Rupert et al., 2004). The first larval stage of bivalves is called a trochophore larva. A ring of cilia around the larvae allow movement through the water. Trochophore larvae do not feed, and lack a digestive system. Instead, they use up yolk energy reserves. After 1 or 2 days the trochophore develops two small transparent shells that are hinged together to enclose the body, and newly developed organs. Larvae at this stage are called veligers. Veligers lack gills but they have simple digestive systems and an organ fringed with cilia (velum) that they can extend out of the shell for swimming and collecting food particles. Particles are then moved towards the mouth by means of shorter cilia (Rupert et al., 2004). These cilia are able to sort and select food of the preferred particle size. In early veligers, particles of 1-2 micrometers in size (*i.e.* small microalgal cells) are preferred but as veligers grow and metamorphose to the next stage (pediveliger) they are able to eat larger phytoplankton (Helm et al., 2004). Once the particles enter the stomach, further sorting occurs and unwanted particles can be passed directly to the intestine for removal. Food particles are digested in the stomach and a digestive gland. After the larval stages, bivalves metamorphose and become juveniles. At this stage, bivalves are morphologically similar to adults but they are smaller and not all their organs are fully developed.

Although microalgae are essential for rearing bivalve molluscs (Helm et al., 2004), they are also very important dietary items for fish and crustacean larvae over the firt larval stages. Thereafter, they actively prey upon small zooplanktonic animals. In hatchery conditions, these live preys are substituted by rotifers and Artemia nauplii (the first non-feeding larval stage) and Artemia metanauplii (second larval stage, usually pre-enriched with microalgae). Such "support" cultures are very labour intensive to run and frequently represent the main variable costs in larviculture rearing operations. The search for nutritional alternatives to live feed holds the promise of developing off-the shelf, nutritionally suitable diets not only for specific species but also designed and manufactured to fulfil the nutritional needs of specific larval stages and substages. The increasing knowledge on larval digestive anatomy and physiology draws closer the possibility of producing efficient larval feeds for valuable marine species.

THE DEMAND FOR LIVE FEED

In general, marine larvae are highly dependent on a constant supply of live food, which is generally provided in hatcheries as microalgae, rotifers and brine shrimp (Artemia).

Globally, the major part of microalgae requirements for aquaculture comes from molluscs for which no inert feed substitution is yet possible (Muller-Feuga, 2000). On the other hand, hatchery production of fish and crustacean larvae requires high amounts of rotifers and Artemia. The production of live food usually accounts for up to 60% of the total variable costs of the commercial hatcheries (Akiyama et al., 1992; Appelbaum, 1989) as these have to produce high volumes of microalgal cultures in order to feed the larvae. In general marine larvae are fed microalgae concentrations ranging from 10 to 200 microalgal cells per microliter, depending on the larval stage, macroalgae cell size and presence of additional feed. Feeding levels are adjusted according to larval density and grazing activity. Under culture conditions, fish and crustacean larvae progressively feed on decreasing algal concentrations until they are weaned onto inert diets, but in contrast, when mollusc larvae metamorphose after setting into the adult form, the concentration of microalgae is slightly increased. Well-established microalgae production protocols are designed to ensure a constant supply of live feed for the mollusc larvae. In addition to live feed, other factors such as salinity, temperature, quality of broodstock, disease management and water quality affect the growth and survival of each larval batch. Although different feeding regimes have been successfully developed to rear marine larvae (most of these for crustacean and fish), the majority include a combination of live and inert feed. Rotifers and Artemia are essential for rearing several species of marine fish and crustaceans. A typical feeding regime for marine fish consists of a microalgae supply followed by transitory combinations of rotifers and Artemia, which are frequently fed to the larvae after they have been nutritionally enriched with microalgae and/or commercial products (usually containing high levels of polyunsaturated fatty acids). Using enriching products allows manipulating the nutritional profile of the zooplanktonic preys in order to better match the nutritional requirements of marine larvae. Although live diets provide a more easily assimilated mixture of soluble proteins, peptides and amino-acids (Fyhn, 1989; Lan and Pan, 1993), and represent advantages such as high digestibility and stability in the water, rotifers and Artemia may lack essential nutrients for marine organisms (Webster and Lovell, 1990; Dhert et al., 2001). Recent research efforts have focused on producing, and nutritionally testing alternative live feeds such as copepods (Camus et al., 2009), cladocerans (Martin et al., 2006), mysid larvae as feed for larval fish (Eusebio et al., 2010) and artificial formulated larval feeds (Conceição et al., 2010, Robinson et al., 2005).

REPLACING LIVE FEED WITH INERT DIETS: A DIFFICULT UNDERTAKING

Depletion of wild stocks, a relatively high market price and the suitability of many marine species for farming have increased the research efforts to increase control on larval production (Fletcher et al., 2007). Enhanced production protocols have effectively led to increased production of fish, crustacean and mollusc larvae, but rearing procedures are still highly dependent on live foods such as microalgae, rotifers and Artemia. The production of live larval feeds demands specific facilities, maintenance expenses, and intensive labour to produce constant amounts of live feeds in a safe and constant fashion (Kanasawa, 1990) so that commercial operations at the hatchery or nursery can keep their economic feasibility. In order to reduce reliance on live feeds research has been conducted over the last decades to find alternatives to live feeds. As previously indicated in this chapter, the production of live feeds is the main factor contributing to the variable costs of for hatcheries as it can represent up to 50% of the total feed costs of postlarvae production (Le Ruyet et al., 1993). Artemia alone can represent up to 40% of the total feed costs, or 80% of the live feed costs (Baskerville-Bridges and Kling, 2000a). Moreover, the nutritive value of planktonic organisms is spatially and temporarily variable, making the use of live food for mass culture restrictive. Therefore, it is necessary to develop larval diets as substitutes for live feeds to further increase the production of healthy larvae and postlarvae for the aquaculture industry (Kanasawa, 1990). Developing effective, species-specific larval diets is a lengthy and resource-demanding process having the final objective of lessening the dependence on the expensive live feeds by ensuring nutritional consistency and product availability. The nutritional transition from live feed to inert feed ("weaning" period) that captive larval and juvenile fish are forced to experience is an extremely important phase in hatcheries (Nhu et al., 2010) and despite recent nutritional and technical advancements, several limitations still exist, including low survival and growth rates, in addition to very high production costs (Baskerville-Bridges and Kling, 2000; Callan et al., 2003).

Crustacean larvae go through rapid and complex larval development, both, morphologically and physiologically. This fact increases the challenge in providing the right type of live and inert feed at specific larval stages. Moreover, accurately assessing incorporation of nutrients into the larval tissue of crustaceans is also a difficult task due to size constrains.

Early crustacean larvae can respond to algal feeding with increased digestive enzyme secretion, but show little ability to respond to the artificial diets (Le Vay et al., 2001). Such a response does develop in later larval stages but is insufficient to produce growth equivalent to that observed with live feeds, which are readily digestible even with low levels of digestive enzyme secretion, in contrast, feeding on the artificial diets increases the digestive enzyme secretion probably as a result from the low availability of dietary protein in the inert feeds due to relatively poor digestibility (Le Vay et al., 1993).

Larvae of marine fish are not the exception in regards to difficulty degree when attempting to substitute live feed with inert diets. Quantifying the processing capacity of the digestive tracts of developing fish larvae for amino acids, fatty acids and other nutrients delivered through live and formulated diets represents a technical challenge, partly because of problems with assessment of ingestion rates and leaching before ingestion (Ronnestad et al., 2007; Southgate and Partridge, 1998).

Research efforts focused on manufacturing formulated diets able to support growth and development of larval fish from first feeding have a clear aim; to reduce to a minimum the need for live organisms in hatcheries (Yufera et al., 2005). A comprehensive understanding of the larval digestive function and processing capacity from the onset of exogenous feeding will provide a better basis for the formulation of larval specific diets for marine species. Constant improvements through nutritional research have narrowed down the dependency on live foods and some studies have reported on the successful total replacement of microalgae and/or Artemia at different larval stages (Kanazawa, 1990; Cuzon and Aquacop, 1998; D'Abramo et al., 2006).

For example, shrimp larval rearing protocols in commercial shrimp hatcheries have been using increasingly higher proportions of inert diets to partially replace live foods in co-feeding regimes (Teshima et al., 2000).

However, feeding only with inert feeds has resulted in poor growth and survival rates (Blair et al., 2003; Stoss et al., 2004). It has been suggested that poor performance of inert diets is due to low residence time in the water (Cahu and Zambonino Infante, 2001; Stoss et al., 2004), low palatability, low ingestion rates due to low digestibility of the diet because of inadequate digestive enzyme activity or poor nutritional composition of the diet (Baskerville-Bridges and Kling, 2000). On the other hand, although live diets represent advantages such as high digestibility and stability in the water, rotifers and Artemia may lack essential nutrients for marine organisms (Léger et al. 1985; Webster and Lovell, 1990; Dhert et al., 2001). In this context, the development of nutritionally complete microdiets that can partially or completely substitute live feed in co-feeding regimes presents clear economic and environmental advantages.

EVALUATION OF LARVAL QUALITY, LARVAL DIETS AND FEEDING PROTOCOLS

Despite extensive research, the quantitative nutritional requirements of larvae of most marine species are not yet well understood and this has been mainly due to the difficulties in quantifying feed intake and assimilation (Le Vay and Gamboa-Delgado, 2011). The latter parameters are typically problematic to estimate in aquatic larval organisms due to size constraints, sample collection difficulties and rapid leaching of nutrients from micro-diets. Two decades ago, Sorgeloos and Leger (1992) described the development of effective finfish and crustacean larval diets for the early first feeding stages as being the major bottleneck for complete replacement of live foods. Despite important advances on the production and nutrition of marine larvae, such statement still holds true after two decades of intensive nutritional research. At the present time, farming techniques for species having high market value (such as the case of yellow fin tuna) have been achieved, but these are currently limited to capturing juvenile fish and growing them in captivity, as larval rearing techniques for this and other species is currently being researched. The optimal performance of a specific larval batch is closely related to the nutritional condition of the captive, broodstock organisms; therefore, a strict monitoring of the rearing conditions and nutrition of the latter is kept. Several types of tests can be conducted to evaluate the larval quality produced by the broodstock and these range from morphological inspections to stress tests (saline and thermal shocks, counter-current swimming activity and immersion of the larval organisms in diverse solutions). Biological parameters such as growth, survival and metamorphosis rates are frequently used as indicators of larval condition (Robinson et al., 2005), while knowledge on the anatomy and physiology of marine larvae greatly assists in developing larval feeds and feeding protocols. Technical aspects such as the effect of water quality parameters, tank shape, size and colour and larval stocking density are frequently used as statistical variables in experimental designs aimed to improve larval rearing conditions.

Evaluation of live diets using different biological parameters is of special importance as some marine larvae respond faster to specific feeding stimulus. For example, some larvae capture their prey only after a visual stimulus triggered by the feeding particle movement, as in the case of zooplanktonic animals swimming. Because of their poor nutritional value and the incapacity of marine fish to elongate or desaturate 18-carbon of longer polyunsaturated fatty acids, it is a common practice to enrich live feeds (Artemia, rotifers) with experimental and commercial products (Haché and Plante, 2011). Live feed is nutritionally manipulated by addition of different emulsions to the culture media, with the ultimate intention of bioencapsulating specific nutrients (fatty acids, vitamins) or therapeutics into the prey, which is used as a vehicle to administer different compounds to the larvae. On the other hand, the nutritional quality of inert feeds has been tested by means of bioassays designed to test the physical presentation of the larval diets which can be manufactured as microparticulate, micro-encapsulated, microbounded and micro-coated particles (Fletcher et al., 2007; Holme et al., 2006; Yufera et al., 2002; Marchetti et al., 1999; Jones et al., 1979) as well as their stability, palatability, digestibility and assimilation. Performance of new larval diets and ingredients are tested by methods such as growth assays, optical and electronic microscopy (Pedroza-Islas et al., 1999), application of inert digestive tracers (Colombo-Hixson et al., 2011), larval amino acid and fatty acid profiling (Conceição et al., 1997; Palacios et al.,

2001), measurement of larval digestive enzyme activity (Kumlu and Jones, 1995), use of radioactive and fluorescent biomarkers (Kelly et al., 2000; Ronnestad et al., 2001) and stable isotope analyses of diets and larvae (Gamboa-Delgado et al., 2008; Jomori et al., 2008; Gamboa-Delgado and Le Vay, 2009). Chemical analyses of diets and animal tissues provide valuable information on the transfer of nutrients from broodstock to egg and from diet to larvae at different stages of a specific species. Continuing research efforts have improved our understanding of how nutrients interact in the larval digestive tract. Current larval feeding regimes for marine fish and crustacean consist of a combination of live and inert feeds. Such feeding protocols take into account the feeding activity and digestive capacity of the larvae at specific developmental stages, the mouth opening size, metamorphosis events and addition of special, immunostimulatory diets for critical periods such as animal transfer and transportation.

FUTURE TRENDS IN MARINE LARVAL REARING

The aquaculture contribution to the world production of the main groups of farmed species has significantly increased since 1950, with the exception of marine fish (FAO, 2010). This fact highlights the intrinsic difficulties in overcoming the challenges posed by the study of new species viable for farming, in this case, the production of marine fish larvae. Nevertheless, it can be easily forecasted that larval culture systems will continue benefitting from the development of new methodologies than in turn, are constantly supported by continued research efforts worldwide. Knowledge on the digestive anatomy and physiology of marine larvae has improved as a result of new, advanced techniques applied to economically important marine species. For example, three-dimensional models of the digestive system during ontogeny have been reconstructed from histological sections of Atlantic cod larvae (Kamisaka and Rønnestad, 2011), generating in this way interactive models showing some outstanding larval anatomical details never reported before: subdivision of digestive tract into foregut, midgut, and hindgut by sphincters; development of stomach and pyloric caeca; location of entrances of bile and pancreatic ducts and ontogeny of pancreas. As important as the anatomical knowledge, findings on physiological aspects such as the ontogenetic programming of digestive enzymes activities will also contribute to develop new larval diets and to enhance larval rearing techniques. Future larval production technologies will take advantage of new developments. For instance, there are interesting examples of intensive, automated larval rearing systems for marine species, which could be upscaled and possibly become common use in the near future (Bosmans et al., 2004; Kolkovski 2004a, 2004b). Management of high to extremely high larval densities cannot be done without paying special attention to critical sanitary aspects as the control and monitoring of bacterial community structures (Nakase, et al., 2007), which also methods tending to automation. Intensifying the larval production in a facility necessarily increases the system complexity and the energy and resource requirements, but such enterprises are still cost-effective as long as the production of larval organisms is high. Most of this production is rapidly demanded by the growing marine aquaculture industry, which is in turn is constantly developing and improving new farming systems on land, near to shore and in open ocean (Su et al., 2000; Kolkovski and Sakakura, 2007; Benetti et al., 2010).

CONCLUSION

The current trend in production yields from aquaculture allows forecasting not only a continued increase in bulk production, but also a diversification of farmed species of crustaceans, molluscs and fish, as a result of new developments in rearing techniques. Once the larval rearing of a specific species is achieved, the critical larval production no longer represents a bottleneck in the whole production process of a particular species. In consequence, there are greater possibilities of successfully breeding this species in captivity through its whole life cycle. Applied research will continue supporting the development of enhanced larval rearing techniques and this will be reflected in higher production of good quality postlarvae and juveniles, which in turn will fulfil the demand of hatchery-produced organisms that the growing aquaculture industry is exerting. The controlled production of phytoplankton needed for the rearing of marine larvae (and also needed for other relatively new applications such as biofuel production) is also increasing at a high rate. Mass production of phytoplankton has been achieved using outdoor "bio-fence" systems supplied with carbon dioxide or using new heterotrophic systems, in which some microalgae species can be grown in dark conditions using a carbon source that replaces the traditional support of light energy for cell photosynthesis. The development of new, species-specific inert larval feeds remains a very challenging task that will probably take decades of intensive nutritional research to effectively lessen the larval dependence on live feed. If this latter aspect is achieved, it will surely represent a major breakthrough in rearing marine larvae and will also contribute to aquaculture diversification as numerous marine species are deemed viable for farming. All these aspects will in conjunction strengthen the "larval rearing" link and in consequence the overall production process, contributing in this way to the remarkable emergence of aquaculture as an important and highly productive activity, which has actually been termed "blue revolution" to compare it with the series of research and technology developments occurring between the 1940s and the late 1970s and which result increased the global agriculture production (green revolution). The appropriate management of marine resources over the next few decades will ultimately indicate if the term "blue revolution" has been well coined.

REFERENCES

Abubakr, M.A., Jones, D.A., 1992. Functional morphology and ultrastructure of the anterior midgut diverticulae of *Penaeus monodon* (Fabricius, 1798) larvae. *Crustaceana* 62, 142-158.

Akiyama D.M., Dominy, W.G., Lawrence., A.L. 1992. Penaeid shrimp nutrition. *In: Marine shrimp culture: principles and practices. Developments in aquaculture and fisheries science.* Vol. 23 (Ed. by A.W. Fast and L.J Lester) Elsevier Science Publisher B.V. The Netherlands. pp. 535-568.

Appelbaum, S., 1989. Can inert diets be used more successfully for feeding larval fish? Thoughts based on indoor feeding behaviour observations. *Polish Archives of Hydrobiology* **36**, 435–437.

Baskerville-Bridges, B., Kling, L.J., 2000. Early weaning of Atlantic cod (*Gadus morhua*) larvae onto a microparticulate diet. *Aquaculture* 189, 109–117.

Benetti, D., 2010. Cobia (*Rachycentron canadum*) hatchery-to-market aquaculture technology: recent advances at the University of Miami Experimental Hatchery (UMEH). *Revista Brasileira de Zootecnia* [online] 39, 60-67.

Benetti, D.D., O'Hanlon, B., Rivera, J.A., Welch, A.W., Maxey, C., Orhun, M.R. 2010. Growth rates of cobia (*Rachycentron canadum*) cultured in open ocean submerged cages in the Caribbean. *Aquaculture* 302, 195–201.

Blair, T., Castell, J., Neil, S., D'Abramo, L., Cahu, C., Harmon, P., Ogunmoye, K., 2003. Evaluation of microdiets versus live feeds on growth, survival and fatty acid composition of larval haddock (*Melanogrammus aeglefinus*). *Aquaculture* 225, 451–461.

Bosmans, J.M.P., Schipp, G.R., Gore, D.J., Jones, B., Vauchez, F., Newman, K.K., 2004. Early weaning of Barramundi, *Lates calcarifer* (Bloch), in a commercial, intensive, semi-automated, recirculated larval rearing system. The Second Hatchery Feeds and Technology Workshop, Sydney, Australia. September 30 – October 1, 2004. http://www.fish.wa.gov.au/docs/op/op030/fop030.pdf.

Cahu, C.L., Zambonino-Infante, J., 2001. Substitution of live food by formulated diets in marine fish larvae. *Aquaculture* 200, 161–180.

Callan C., Jordaan A., Kling L.J., 2003. Reducing Artemia use in the culture of Atlantic cod (*Gadus morhua*). *Aquaculture* 219, 585-595.

Camus, T., Zeng, Ch., McKinnon, A.D., 2009. Egg production, egg hatching success and population increase of the tropical paracalanid copepod, *Bestiolina similis* (Calanoida: Paracalanidae) fed different microalgal diets. *Aquaculture* 297, 169-175.

Colombo-Hixson, S.M., Olsen, R.E., Milley J.E., Lall, S.P., 2011. Lipid and fatty acid digestibility in Calanus copepod and krill oil by Atlantic halibut (*Hippoglossus hippoglossus* L.). *Aquaculture* 313, 115-122.

Conceição L.E.C., van der Meeren, T., Verreth J.A.J., Evjen, M.S., Houlihan, D.F., Fyhn, H.J., 1997. Amino acid metabolism and protein turnover in larval turbot (*Scophthalmus maximus*) fed natural zooplankton or Artemia. *Marine Biology* 129, 255-265.

Conceição, L.E.C., Yúfera, M., Makridis, P., Morais, S., Dinis, M.T., 2010. Live feeds for early stages of fish rearing. *Aquaculture Research* 41, 613-640.

Cuzon, G., Aquacop., 1998. Nutritional review of *Penaeus stylirostris*. In Reviews in Fisheries Science 6 (Stickney, R. ed.) CRC Press. 129-141 pp.

D'Abramo, L.R., Perez, E.I., Sangha, R., Puello-Cruz, A., 2006. Successful culture of larvae of *Litopenaeus vannamei* fed a microbound formulated diet exclusively from either stage PZ2 or M1 to PL1. *Aquaculture* 261, 1356–1362.

Delmendo, M.N., 1990. Bivalve farming: An alternative economic activity for small-scale coastal fishermen. ASEAN/UNDP/FAO Regional Small-Scale Coastal Fisheries Development Project, Manila, Philippines. ASEAN/SF/89/Tech. 11. 45 p.

Dhert, P., Rombaut, G., Suantika, G., Sorgeloos, P., 2001. Advancement of rotifer culture and manipulation techniques in Europe. *Aquaculture* 200, 129–146.

Eusebio, P.S., Coloso, R.M., Gapasin, R.S.J., 2010. Nutritional evaluation of mysids *Mesopodopsis orientalis* (Crustacea: Mysida) as live food for grouper *Epinephelus fuscoguttatus* larvae. *Aquaculture* 306, 289-294.

FAO, 2010. The state of the world fisheries and aquaculture (SOFIA) 2010. FAO, *Fisheries and Aquaculture Department*, Rome, Italy, 242 pp.

Fletcher, R.C., Roy, W., Davie, A., Taylor, J., Robertson, D., Migaud, H., 2007. Evaluation of new microparticulate diets for early weaning of Atlantic cod (*Gadus morhua*): Implications on larval performances and tank hygiene. *Aquaculture* 263, 35–51.

Fyhn, H.J., 1989. First feeding of marine fish larvae: are free amino-acids the source of energy? *Aquaculture* 80, 111-120.

Gamboa-Delgado, J., Cañavate, J.P., Zerolo, R., Le Vay, L. 2008. Natural carbon stable isotope ratios as indicators of the relative contribution of live and inert diets to growth in larval Senegalese sole (*Solea senegalensis*). *Aquaculture* 280, 190-197.

Gamboa-Delgado, J., Le Vay, L., 2009. *Artemia* replacement in co-feeding regimes for mysis and postlarval stages of *Litopenaeus vannamei*: Nutritional contribution of inert diets to tissue growth as indicated by natural carbon stable isotopes. *Aquaculture* 297, 128-135.

Haché, R., Plante, S., 2011. The relationship between enrichment, fatty acid profiles and bacterial load in cultured rotifers (Brachionus plicatilis L-strain) and Artemia (Artemia salina strain Franciscana. *Aquaculture* 311, 201-208.

Helm, M.M., Bourne, N., Lovatelli, A., 2004. Hatchery culture of bivalves. A practical manual. Helm, M.M.; Bourne, N.; Lovatelli, (comp/ed). FAO Fisheries Technical Paper. No. 471. Rome, FAO. 177p. http://www.fao.org/docrep/007/y5720e/y5720e00.htm.

Holme, M., Zeng, C., Southgate, P.C., 2006. Use of microbound diets in larval culture of the mud crab, *Scylla serrata*. *Aquaculture* 257, 482-490.

Jomori, R.K., Ducatti, C., Carneiro, D.J., Portella, M.C., 2008. Stable carbon (δ^{13}C) and nitrogen (δ^{15}N) isotopes as natural indicators of live and dry food in *Piaractus mesopotamicus* (Holmberg, 1887) larval tissue. *Aquaculture Research* 39, 370–381.

Jones, D.A., Kamarudin, M.S., Le Vay, L., 1993. The potential for replacement of live feeds in larval culture. *Journal of the World Aquaculture Society* 24, 199-210.

Jones, D.A., Kanazawa, A., Abdel-Rahman, S., 1979. Studies on the presentation of artificial diets for rearing the larvae of *Penaeus japonicus* Bate. *Aquaculture* 17, 33-43.

Jones, D.A., Kumlu, M., Le Vay, L., Fletcher, D.J., 1997. The digestive physiology of herbivorous, omnivorous and carnivorous crustacean larvae: a review. *Aquaculture* 155, 285-295.

Kamarudin, M.S., 1992. Studies on the digestive physiology of crustacean larvae. PhD thesis. University of Wales, Bangor, UK.221 pp.

Kamisaka, Y., Rønnestad, I., 2011 Reconstructed 3D models of digestive organs of developing Atlantic cod (*Gadus morhua*) larvae. *Marine Biology* 158, 233-243.

Kanazawa, A., 1990. Microparticulate feeds for Penaeid larvae. Advances in Tropical Aquaculture Workshop, Feb 20-Mar 4, 1989, Tahiti, French Polynesia, *AQUACOP IFREMER, Actes de Colloque* 9, 395–404.

Kelly, S.P., Larsen, S.D., Collins, P.M., Woo, N.Y.S., 2000. Quantitation of inert feed ingestion in larval sea bream (*Sparus sarba*) using auto-fluorescence of alginate-based microparticulate diets. *Fish Physiology and Biochemistry* 22, 109-117.

Kolkovski, A., Sakakura, Y., 2007. Yellowtail Kingfish culture — opportunities and problems. *World Aquaculture* 38, 7–13.

Kolkovski, S., Curnow, J., King, J., 2004a. Intensive rearing system for fish larvae research – I. marine fish larvae rearing system. *Aquaculture Engineering* 31, 295-308.

Kolkovski, S., Curnow, J., King, J., 2004b. Intensive rearing system for fish larvae research – II. Artemia hatching and enriching system. *Aquaculture Engineering* 31:309-317.

Koven, W., Kolkovski, S., Hadas, E., Gamsiz, K., Tandler, A., 2001. Advances in the development of microdiets for gilthead seabream, *Sparus aurata*: a review. *Aquaculture* 194, 107–121.

Kumlu, M., Jones, D.A., 1995. The effect of live and artificial diets on growth, survival and trypsin activity in larvae *Penaeus indicus*. *Journal of the World Aquaculture Society* 26, 406–415.

Lan, C.C., Pan, B.S., 1993. In-vitro digestibility simulating the proteolysis of feed protein in the midgut gland of grass shrimp (*Penaeus monodon*). *Aquaculture* 109, 59– 70.

Le Ruyet, J.P., Alexandre, J.C., Thebaud, L., Mugnier, C., 1993. Marine fish larvae feeding: formulated diets or live prey? *Journal of the World Aquaculture Society* 24, 211–224.

Le Vay, L., Gamboa-Delgado, J., 2011. Naturally-occurring stable isotopes as direct measures of larval feeding efficiency, nutrient incorporation and turnover. Larvi ´09 Special Issue. *Aquaculture* 315, 95-103.

Le Vay, L., Jones, D.A., Puello-Cruz, A C., Sangha, R.S., Ngamphongsai, C., 2001. Digestion in relation to feeding strategies exhibited by crustacean larvae. *Comparative Biochemistry and Physiology A* 128, 623–630.

Le Vay, L., Rodrıguez, A., Kamarudin, M., Jones, D.A., 1993. Influence of live and artificial diets on tissue composition and trypsin activity in *Penaeus japonicus* larvae. *Aquaculture* 118, 287–297.

Léger, P., Bieber, G.F., Sorgeloos, P., 1985. International study on Artemia. XXXIII. Promising results in larval rearing of *Penaeus stylirostris* using a prepared diet as algal substitute and for *Artemia* enrichment. *Journal of the World Aquaculture Society* 16, 354– 367.

Lovett, D.L., Felder, D.L., 1990a. Ontogenetic change in enzyme distribution and midgut function in developmental stages of *Penaeus setiferus* (Crustacea Decapoda Penaeidae). *Biology Bulletin* 178, 144-159.

Lovett, D.L., Felder, D.L., 1990b. Ontogenetic change in enzyme distribution and midgut function in developmental stages of *Penaeus setiferus* Crustacea Decapoda Penaeidae. *Biology Bulletin* 178, 160-174.

Marchetti, M., Tossani N., Marchetti, S., and Bauce, G. 1999. Stability of crystalline and coated vitamins during manufacture and storage of fish feeds. *Aquaculture Nutrition* 5, 115-120.

Martin, L., Arenal, A., Fajardo, J., Pimentel, E., Hidalgo, L., Pacheco, M., Garcia, C., Santiesteban, N., 2006. Complete and partial replacement of Artemia nauplii by *Moina micrura* during early postlarval culture of white shrimp (*Litopenaeus schmitti*). *Aquaculture Nutrition* 12, 89-96.

McDonald, N.L., Stark, J.R., Keith, M., 1989. Digestion and nutrition of the prawn *Penaeus monodon*. *Journal of the World Aquaculture Society* 20: p. 53A.

Moss, S., Arce, S., 2003. Disease Prevention Strategies for Penaeid Shrimp Culture. *Pathobiology and Aquaculture of Crustaceans*, (Rosenberry), 1-12.

Muller-Feuga, A., 2000. The role of microalgae in aquaculture: situation and trends. *Journal of Applied Phycology* 12, 527-534.

Nakase, G., Nakagawa, Y., Miyashita, S., Nasu, T., Senoo, S., Matsubara, H., Eguchi, M., 2007. Association between bacterial community structures and mortality of fish larvae in intensive rearing systems. *Fisheries Science* 73, 784–791.

Nhu, V., Dierckens, K., Nguyen, H., Hoang, T. T., Le, T., Tran, M., Nys, C., 2010. Effect of early co-feeding and different weaning diets on the performance of cobia (*Rachycentron canadum*) larvae and juveniles. *Aquaculture* 305, 52-58.

Palacios, E., Racotta, I.S., Heras, H., Marty, Y., Moal, J., Samain, J.F., 2001. The relation between lipid and fatty acid composition of eggs and larval survival in white pacific shrimp (*Penaeus vannamei*, Boone, 1931). *Aquaculture International* 9, 531– 543.

Pedroza-Islas, R. Vernon-Carter, E.J., Durán, D.C., Trejo-Martínez, S., 1999.Using biopolymer blends for shrimp feedstuff microencapsulation. I. Microcapsule particle size, morphology and microstructure. *Food Research International* 32, 367-374.

Robinson, C.B., Samocha, T.M., Fox, J.M., Gandy, R.L., McKee, D.A., 2005. The use of inert artificial commercial food sources as replacements of traditional live food items in the culture of larval shrimp, *Farfantepenaeus aztecus*. *Aquaculture* 245, 135-147.

Ronnestad, I., Kamisaka, Y., Conceição, L. C., Morais, S., Tonheim, S., 2007. Digestive physiology of marine fish larvae: Hormonal control and processing capacity for proteins, peptides and amino acids. *Aquaculture* 268, 82-97.

Rønnestad, I., Rojas-García, C.R., Tonheim, S.K., Conceição, L.E.C., 2001. In vivo studies of digestion and nutrient assimilation in marine fish larvae. *Aquaculture* 201, 161–175.

Ruppert, E.E., Fox, R.S., Barnes, R.D., 2004. Invertebrate zoology: a functional evolutionary approach. Belmont, CA: Thomas-Brooks/Cole.963 pp.

Sorgeloos, P., Leger, P., 1992. Improved larviculture outputs of marine fish, shrimp and prawn. *Journal of the World Aquaculture Society* **23,** 251–264.

Southgate, P.C., Partridge, G.J., 1998. Development of artificial diets for marine finfish larvae: problems and prospects. *In*: *Tropical Mariculture* (De Silva, S.S. ed.), Academic Press, London, U.K. pp. 151– 170.

Stoss, J., Hamre, K., Ottera, H., 2004. Weaning and nursery. *In*: Moksness, E., Kjorsvik, E., Olsen, Y. (Eds.), *Culture of Cold- Water Maine Fish*. Fishing News Books, Blackwell Publishing, Oxford, U.K.

Su, M.S., Chien, Y.H., Liao, I.C., 2000. Potential of marine cage aquaculture in Taiwan: cobia culture. In: Liao, I.C., Lin, C.K. (Eds.), Cage Aquaculture in Asia: Proceedings of the First international Symposiumon Cage Aquaculture in Asia, Asian Fisheries Society, Manila, and World Aquaculture Society — Southeast Asian Chapter, Bangkok, pp. 97– 106.

Teshima, S., Ishikawa, M., Koshio, S., 2000. Nutritional assessment and feed intake of microparticulate diets in crustaceans and fish. *Aquaculture Research* 31, 691-702.

Tucker, J.W., 1998, Marine Fish Culture. *Kluwer Academic Publishers*, MS, USA. 750 pp.

Utting, S.D., Spencer, B.E., 1991. The hatchery culture of bivalve mollusc larvae and juveniles. Lab. Leafl., *MAFF Fish. Res., Lowestoft*, No 68: 31 pp.

Webster, C.D., Lovell, R.T., 1990. Quality evaluation of four sources of brine shrimp Artemia spp. *Journal of the World Aquaculture Society* 21, 180–185.

Yufera, M., Fernandez-Diaz, C., Pascual, E., 2005. Food microparticles for larval fish prepared by internal gelation. *Aquaculture* 248, 253-262.

Yufera, M., Kolkovski, S., Fernandez-Diaz, C., Dabrowski, K., 2002. Free amino acid leaching from a protein-walled microencapsulated diet for fish larvae. *Aquaculture* 214, 273–287.

In: Larvae: Morphology, Biology and Life Cycle
Editors: Kia Pourali and Vafa Niroomand Raad

ISBN: 978-1-61942-662-7
© 2012 Nova Science Publishers, Inc.

Chapter 7

MORPHOLOGY LIKE A TOOL TO UNDERSTAND THE SOCIAL BEHAVIOR IN ANT'S LARVAE

Gabriela Ortiz, Maria Izabel Camargo Mathias and Odair Correa Bueno*

Universidade Estadual Paulista Júlio de Mesquita filho (UNESP),
Instituto de Biociências, Departamento de Biologia, Rio Claro/SP, Brasil

ABSTRACT

All metamorphosing insects pass through four consecutive stages before reaching their adult form and these stages are known as the egg, the larva, the semipupa and the pupa. The larval stage comprising individuals who have completely different morphology of the other three, since the body is devoid of external appendages and with internally adipose tissue. From the larval stage begins the growth of individuals who shed their skins several times and depending on the insect considered the substages vary from 3 to 5. In bees in general there are five larval stages called L1 to L5, interspersed with each other for four periods of change ending in larval and pupal changes when successful-L5 larvae enter the pupal stage, where individuals have had adult characteristics. Since ants have about four larval stages (L1, L2, L3, L4), but this number can vary from 3 to 6 depending on the species.

In the larvae of ants there are morphological peculiarities in certain species that set them apart from most reported in the literature so far. An interesting fact was observed in the Argentine ant *Linepithema humile*, an invasive species, since they present a dorsal protuberance in the sixth abdominal segment found exclusively in their larvae. Researchers have tried to explain the social function of this structure. One function described was the nutrition. However the morphological results obtained in this study suggested that the dorsal protuberance has no secretory function, since ducts or pore channels were not found, only small folds in the cuticle, which does not indicate a secretion release function. The secretion released is just for the cuticle compounds and does not synthesize other one which would be important in social behavior. Therefore the unique social function that can be attributed to the dorsal protuberance was mechanical or

---

* Email: gaortiz@rc.unesp.br.

serving like a facilitator making it easier for the workers to transport the immatures through the different environments in the colony.

Another relevant fact would be peculiar behavior observed in basal attine ants, which in addition to cultivating their fungus, still keeps the fungus growing externally on the surface of the cuticle's larvae, behavior not seen in the genre Atta. Therefore was verified the alterations in the cuticle of the larvae from *Myrmicocrypta*, *Mycetarotes* and *Trachymyrmex* and this structure is formed by a simple cubic epithelium, whose cells possibly change its shape to prismatic, depending on their secretory activity. The presence of fungi hyphae was observed both on the external side of the basal attine larvae as well as emitting projections to the interior of the cuticle reaching the epithelium and the adipocyte cells. Some authors have suggested that the fungus could benefit the larvae protecting them against occasional pathogens by forming a physical or chemical barrier. Data obtained in this study demonstrated that the fungus deposited on the surface of immature from basal attine maintain a close relationship with them, once the fungus hyphae have the ability to disorganize the lamellar cuticle, penetrating the interior of the cells through the emission of prolongations.

INTRODUCTION

Ants are social insects of the family Formicidae and, along with the related wasps and bees, belong to the order Hymenoptera. Ants evolved from wasp-like ancestors in the mid-Cretaceous period between 110 and 130 million years ago and diversified after the rise of flowering plants. More than 12,500 out of an estimated total of 22,000 species have been classified. They are easily identified by their elbowed antennae and a distinctive node-like structure that forms a slender waist (Hölldobler and Wilson, 1990).

Ants form colonies that range in size from a few dozen predatory individuals living in small natural cavities to highly organized colonies which may occupy large territories and consist of millions of individuals. These larger colonies consist mostly of sterile wingless females forming castes of "workers", "soldiers", or other specialized groups. Nearly all ant colonies also have some fertile males called "drones" and one or more fertile females called "queens". The colonies are sometimes described as superorganisms because the ants appear to operate as a unified entity, collectively working together to support the colony (Hölldobler and Wilson, 1990).

Ants have colonized almost every landmass on Earth. The only places lacking indigenous ants are Antarctica and a few remote or inhospitable islands. Ants thrive in most ecosystems, and may form 15–25% of the terrestrial animal biomass. Their success in so many environments has been attributed to their social organisation and their ability to modify habitats, tap resources, and defend themselves. Their long co-evolution with other species has led to mimetic, commensal, parasitic, and mutualistic relationships (Hölldobler and Wilson, 1990).

The life of an ant starts from an egg. If the egg is fertilized, the progeny will be female (diploid); if not, it will be male (haploid). Ants develop by complete metamorphosis with the larval stages passing through a pupal stage before emerging as an adult. The larva is largely immobile and is fed and cared for by workers. Food is given to the larvae by trophallaxis, a process in which an ant regurgitates liquid food held in its crop. This is also how adults share food, stored in the "social stomach", among themselves. Larvae may also be provided with solid food such as trophic eggs, pieces of prey and seeds brought back by foraging workers and may even be transported directly to captured prey in some species (Hölldobler and Wilson, 1990).

All metamorphosing insects pass through four consecutive stages before reaching their adult form and these stages are known as the egg, the larva, the semipupa and the pupa. The larval stage comprising individuals who have completely different morphology of the other three, since the body is devoid of external appendages and with internally adipose tissue. From the larval stage begins the growth of individuals who shed their skins several times and depending on the insect considered the substages vary from 3 to 5. In bees in general there are five larval stages called L1 to L5, interspersed with each other for four periods of change ending in larval and pupal changes when successful-L5 larvae enter the pupal stage, where individuals have had adult characteristics. Since ants have about four larval stages (L1, L2, L3, L4), but this number can vary from 3 to 6 depending on the species.

In the larvae of ants there are morphological peculiarities in certain species that set them apart from most reported in the literature so far. An interesting fact was observed in the Argentine ant *Linepithema humile*, an invasive species, since they present a dorsal protuberance found exclusively in their larvae. The biological success of the argentine ant as invasive species is closely related to the size of its colonies, formed by groups of individuals that interact in a cooperative way (Pedersen et al., 2006). For most species, the social structure has been described as multicolonial, where one colony is formed by a nest populated by workers and queens. As for *L. humile*, invasive populations from other regions as well as local populations are unicolonial; therefore, different nests can be found in the same colony. In this kind of social structure, the workers from different nests do not compete for resources and live together, characterizing a supercolony, which can be composed of hundreds of nests that host million of individuals (Giraud et al., 2002; Queller, 2000; Tsutsui and Suarez, 2003; Tsutsui et al., 2000).

New information about the behavior concerning offspring care in ant colonies is important to understand the social organization and work division of these insects. The argentine ant *L. humile* curiously presents different larvae, i.e., they present a dorsal protuberance in the sixth abdominal segment. Information about the histological organization and social function of this protuberance remain scant because the major interest of the researchers is focused on the adults, aiming at controlling these insects; once this ant species is one of the most economically important because of the high levels of infestation they are able to reach. This structure presents short and scarce hairs cover the larvae; however, they are not present over the surface of the protuberance and did not presents pores or other kind of cuticular opening that could indicate the release of any kind of secretion.

Another relevant fact would be peculiar behavior observed in basal attine ants, which in addition to cultivating their fungus, still keeps the fungus growing externally on the surface of the cuticle's larvae, behavior not seen in the genre Atta. The attine phylogeny is complicated and the *Myrmicocrypta*, *Mycetarotes* and *Trachymyrmex* genera have little in common with one another except that species in all the genera cultivate fungus on their larvae. In this way they differ from *Atta* genus. For convenience, we'll refer to the first three genera (*Myrmicocrypta*, *Mycetarotes* and *Trachymyrmex*) as "basal" and to *Atta* as "derived", although these terms have little meaning (*Myrmicocrypta* may be high derived in its own unique way) except to distinguish *Atta* from the remaining three.

Murakami and Higashi (1997) and Lopes et al. (2005) provide the most detailed description to date of the behavior of fungus-planting on attine brood, noting that this behavior is indistinguishable from the fungus-planting behavior on garden substrate during normal gardening. In essence, workers treat the developing brood somewhat like garden substrat, inserting the brood into the garden matrix, then planting mycelial tufts onto the brood with the same kind of tamping motions of the front legs that typify normal fungiculture (Weber, 1972; Murakami and Higashi, 1997; Lopes et al., 2005). Despite the larger size of male larvae, planting of fungal tufts occurs more frequently onto worker larvae than onto male larvae of *Acromyrmex subterraneus* (Camargo et al., 2006). In *Acromyrmex* ants, larvae are first thoroughly licked by workers before they plant mycelium onto the larval integument, but workers cease licking after planting (Lopes et al., 2005). Mycelium planted on gardening substrate is fertilized initially with fecal droplets of workers (Weber, 1972), but mycelium planted on larvae is not tended in any comparable way by workers (Lopes et al., 2005), suggesting that the mycelium growing on larvae may subsist in part on secretions derived from the larval integument. Fungus-coating of brood varies between attine species. For example, brood of species in *the Apterostigma pilosum* species group (Lattke, 1997; Villesen et al., 2004) are covered with a thick coat, and the fungus-covered brood is integrated semi-permanently into the garden matrix. In contrast, fungus-coating of brood appears completely absent in the genus *Atta* and in several, phylogenetically disparate, lower attine species. The mycelial coat of brood is also present in the yeast cultivating *Cyphomyrmex* species that grow their cultivars in a non-mycelial, yeast state (Wheeler, 1907; Mueller et al., 1998; Wang et al., 1999), but the same fungus is thought to grow as a mycelium on the integument of brood of the yeast-cultivating *Cyphomyrmex* species (Schultz and Meier, 1995).

Biological studies have revealed how complex the symbiotic association is between ants and fungus. The discovery of *Pseudonocardia* bacteria as a new symbiont, which grows on the integument of attine ants and produces antibiotics that may help to control the *Escovopsis* pathogen, as well as studies of the relationship between the participants of the symbiosis and the obtaining of new information, could foment new lines of research which might increase the knowledge about this association, unique in nature (Currie et al., 2003). The confirmation of this symbiosis is the odd association established between basal ants from the tribe Attini with their fungus, since they use the surface of their own offspring to host them, usually depositing them on the eggs, larvae and pupae. Within the ant ontogenesis the larval stage is marked by a cycle of molts with the periodic substitution of the cuticle, the period when the individual effectively grows. After the larva reaches its maximum size metamorphosis takes place with the formation of a new cuticle, resistant enough to allow the survival of the adult insect in a new environment. The molting process begins in the apolysis, with the separation of the epidermal cells of the old cuticle via molt fluid secretion which contains proteases and chitinases enzymes, which digest the main constituents of the old cuticle (Reynolds and Samuels, 1996). The insect cuticle are formed by a series of layers made up of chitin and proteins, in addition to lipids, salts and pigments, and represent a temporal register of the epidermal cells synthesis cycle (Wigglesworth, 1948, 1957; Richards, 1951; Hackman, 1971; Hepburn, 1985).

The cuticle of workers from basal genera *Myrmicocrypta* (Paleoattini clade), *Mycetarotes* (basal group from Neoattini clade) and *Trachymyrmex* (transitional group from Neoattini clade) have the habit of depositing small fragments of fungus on the larvae cuticle, characterizing an unusual behavior in the social relationship of leaf-cutting ants. Histological techniques were applied, allowing the observation of possible alterations in the cuticle of larvae from the studied species, including the insertion of the fungus in the interior of the larvae, establishing a new type of symbioses between ant and fungus.

Some general aspects of the insect cuticle organization have been established through light microscopy studies, while electronic scanning has provided information about the macromolecular structure of the cuticle detailing its surface and reporting its relation to the epidermal cells which secreted it (Smith, 1968). The chemical composition of the insect cuticle was previously described by Hackman (1964), who found that exo and endocuticular layers were composed of chitin polysaccharide associated with protein. For a deeper study of the insects' integument, morphological techniques were used, indispensable tools for the understanding of the tissue and cellular organization of the organisms. In this study were applied histological, scanning electron microscopy and transmission electron microscopy techniques, bringing information about these structures.

RESULTS AND DISCUSSION

SEM allowed the observation of the presence of dorsal protuberance in *L. humile* larvae, typical morphological characteristic of immature *Linepithema* (Figures 1A, B). It is placed in the ninth segment of the larva, which corresponds to the sixth abdominal segment. Short and scarce hairs cover the larvae; however, they are not present over the surface of the protuberance (Figure 1B).

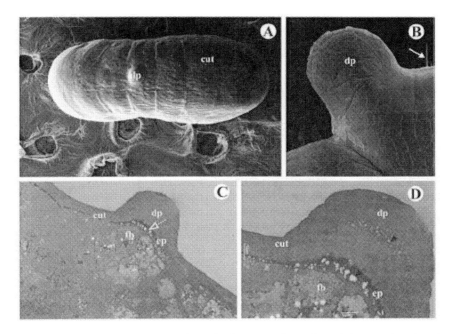

Figure 1. A. Scanning electron microscopy (SEM) of a *Linepithema humile* larvae, showing the presence of dorsal protuberance (dp) located in the sixth abdominal segment. Scale bar = 300μm. B. Detail of the dorsal protuberance (dp) where there are hairs (arrow). Scale bar = 20 μm. C.D. Semithin sections of epidermis stained with Azur II and methylene blue, where was observed the dorsal protuberance region (dp); the cuticle (cut), thicker in this region; the prismatic epithelium (ep); and the vacuoles (arrow head) as well as fat body cells (fb). C.D. Scale bar = 10μm.

SEM of the protuberance did not detect the presence of pores or other kind of cuticular opening that could indicate the release of any kind of secretion. The results showed that their integument is composed of external cuticle, an epidermis composed of a single epithelium, and the existence of fat body cells supporting this structure, and these cells are thicker and more evident in the dorsal protuberance region (Figures 1; 2, 3, 4, 5). The second layer of the cuticle was actually an acellular region (as well as the cuticle itself), but composed of granulous-like not lamellar material (Figures 2B, 2D; 3A).

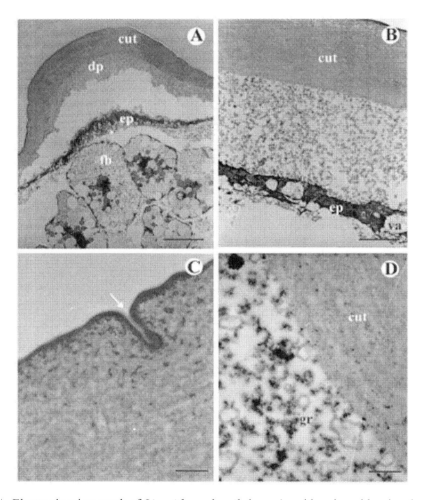

Figure 2. A. Electronic micrograph of *Linepithema humile* larvae's epidermis, evidencing the dorsal protuberance region (dp), where was observed the thickened cuticle (cut) and epithelium (ep), in addition to the increase in the number of fat body cells (fb) in the protuberance region. Scale bar = 50 μm. B. Detail of the cuticle (cut) and flattened cells of the epithelium (ep), with huge vacuoles (va). Scale bar = 5μm. C. Detail of the cuticle, where there is a fold (arrow) next to the dorsal protuberance region. Scale bar = 1μm. D. Detail of the cuticle limit (cut)/granular region (gr) in the dorsal protuberance region. Scale bar = 1μm.

This granular region has direct contact with the flattened epidermic cells (over all the larva's extension) and also with the prismatic ones (dorsal protuberance region) (Figures 2B, 2D; 3A). According to Rossini (1997), who studied meliponinae bees larvae, this acellular region could be a place for the accumulation of molt fluid between epidermis and cuticle, where the granules containing chitinases and proteinases, enzymes responsible for the digestion of the old cuticle could be frequently observed.

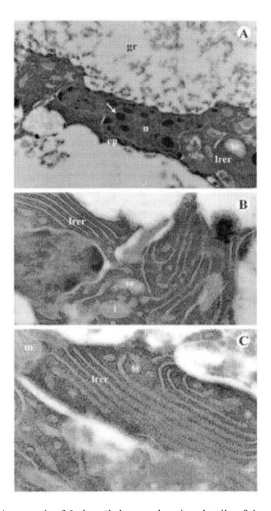

Figure 3. A. Electronic micrograph of *L. humile* larvae showing details of the epithelium flattened cells (ep) with flat nuclei with clot shaped chromatin (arrow) and granular region (gr), in direct contact with the epithelium. There are cisterns of the lamellar rough endoplasmic reticulum (lrer) with secretion (se) in its interior. Scale bar = 2μm. B.C. Details of flattened cells cytoplasm where there is a great amount of lamellar rough endoplasmic reticulum (lrer), containing secretion droplets (se) in the interior of the cisterns. There are lipid droplets (l) and mitochondria (m) in these cells. B.C: Scale bar = 1μm.

Flattened epidermic cells in extradorsal was observed that in *L. humile* not only the prismatic cells are secretory, but also all the epidermic cells, including the flattened ones. This function would be proved by the existence of extended lamellar rough endoplasmic reticulum filling large extensions of the cytoplasm in all the epidermis extension, assuring that these cells would be actively involved with the production of substances that would be exported to the surface of the animal's body (Figures 3A-C). The flattened cells, also the prismatic ones presented only in dorsal protuberance region had extended microvilli both in the surfaces turned to the cuticular layer (apical) and in the fat body cells layer (basal) (Figures 4B-D). These microvilli presented secretion droplets that could be exocyted in both directions, acting like a transport to both directions.

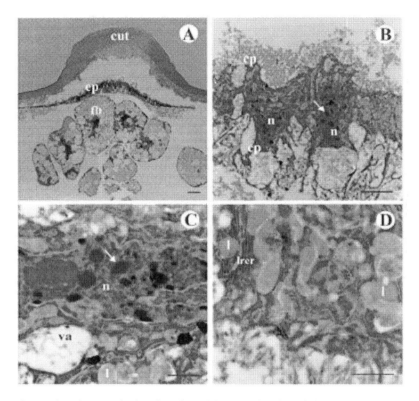

Figure 4. A. Electronic micrograph showing the epidermis of *L. humile* larvae in the dorsal protuberance region, where was observed the cuticle (cut), thickened epithelium prismatic cells (ep), as well as fat body cells (fb). Scale bar = 3 μm. B. Detail of the prismatic cells with star-shaped nucleus (n) from where was observed cytoplasmic prolongations (cp), similar to microvilosities toward the cuticular layer and subjacent fat body cells. Scale bar = 3μm. C. Detail of the flattened cells, where the nucleus (n) is evident and electron-dense clots (arrow) are present. There are vacuoles (va) and lipid droplets (l) in these cells. Scale bar = 1μm. D. Detail of the basal domain of prismatic cells, where there is lamellar rough endoplasmic reticulum (lrer), as well as secretion droplets and lipid droplets (l). Scale bar = 1μm.

According to Rossini (1997), the presence of microvilli in the surface of the insect's epidermic cells would be an excellent indicator of the occurrence of secretory processes, mainly in the periods of molt and intermolt, when they would reach maximum development during the molt period, decreasing gradually in the intermolt period. *L. humile* larvae epidermic cells showed strong adherence through extense interdigitation of the plasmatic membrane, which allowed a significant increase in the contact surface among them. This kind of adherence is very common in the epithelium cells once this is a friction-prone region (Junqueira and Carneiro, 2004). Several mitochondria with varied shape and size were also observed with significant frequency and varied morphology in the cytoplasm of the epidermic cells (flattened and prismatic) of *L. humile* larvae (Figures 3B, C), suggesting an intense demand of energy for the performance of physiologic processes in these insects. Other characteristic of the *L. humile* larvae integument was the great number of electronlucid secretion droplets, probably with lipidic content arising from the epidermic cells to the cuticle, indicating once more the participation of them in the process of secretion of substances that constitute the cuticle (Figures 5B, C). *L. humile* larvae fat body cells filling the space subjacent to epidermic cells were big, and their cytoplasm contained lipidic droplets (electron-lucid) and granules with medium to high electron density, suggesting its protein nature or even suggesting that these could be the result of complexes between different elements (lipids, proteins, and carbohydrates) (Figure 5).

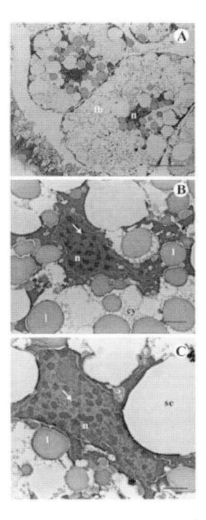

Figure 5. A. Electronic micrograph detailing fat body cells structure (fb) adjacent to the epidermis, where there are star-shaped nuclei (n). Scale bar = 20μm. B.C. Details of the fat body cells cytoplasm (cy) where there are star-shaped nuclei (n) and chromatin clots (arrow). There are also electron-lucid lipid droplets (l) and secretion (se) with electron density varying from medium to strong. B. Scale bar = 3μm. C. Scale bar = 2μm.

The secretory role of the fat body cells has already been registered by other authors, and today it is known that in addition to the lipids these cells have great ability to synthesize and store protein granules that could be eventually used when the insect faced a stressing situation, like starvation (Roma et al., 2010).

More recent morphologic studies addressing aspects of *L. humile* larvae only mentioned and described *Linepithema* dorsal protuberance externally and referred to it as a characteristic that differentiates the immature forms of the genus, considering that this structure is only present in the female larvae (Shattuck, 1992; Solis et al., 2010).

The observation of histological and ultrastructural sections allowed to establish the distribution of the cells and tissues in the epidermal region of *Atta sexdens rubropilosa* (control) and *Mycetarotes parallelus*, *Trachymyrmex fuscus* and *Myrmicocrypta* sp. ant larvae (Figures 6, 7, 8, 9).

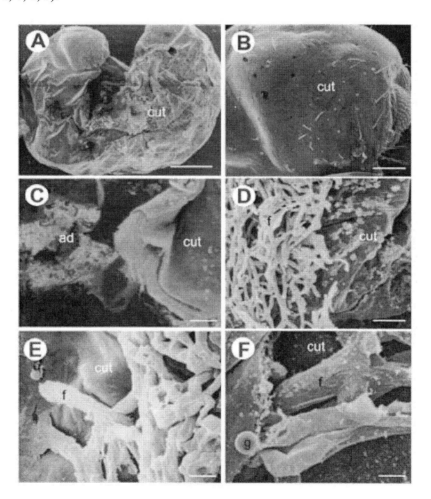

Figure 6. A. B. General view of the larvae of *A. sexdens rubropilosa*. A. Bar = 250µm. B. Bar = 100µm. C. Detail of the cuticle and adipose cells. Scale bar = 100µm. D. Details of the fungus covering the surface of the larvae of *M. parallelus*. Scale bar = 50µm. E. Details of the fungus covering the surface of the larvae of *Myrmicocrypta* sp. Scale bar = 10µm. F. Details of the fungus covering the surface of the larvae of *T. fuscus*. Scale bar = 10µm: cut = cuticle, ad = adipose cells, f = fungi hyphae, g= gongylidea.

The application of techniques of scanning electron microscopy shows the surface of the larvae of *Atta sexdens rubropilosa* (control) and *Mycetarotes parallelus*, *Myrmecocrypta* and *Trachymyrmex fuscus* sp. (Figure 6), which covering the larvae of *A. sexdens ruprobilosa* can observe the presence of short and sparse bristles, which are found in specific regions of the body, i.e., preferentially located in the head and anus (Figure 6B). The surface of the cuticle of these larvae, and the bristles, also has channels and pores that are located irregularly distributed along its entire length. In some specimens of *A. sexdens ruprobilosa* analyzed, which had broken its cuticle, one can observe the presence of adipose cells filling the subepidermal space (Figure 6C).

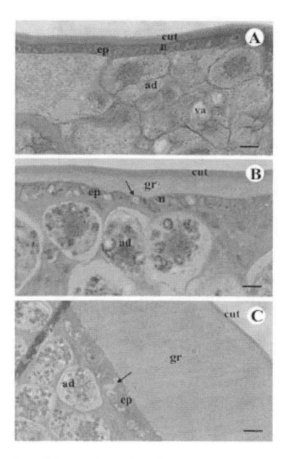

Figure 7. Histological sections of *Atta sexdens rubropilosa* ants represent successively more advanced degrees of separation of the cuticle during molting stained with Hematoxylin–Eosin. A. Detail of simple cubic epithelial cells (ep) with roundshaped nucleuses (n) and of the cuticle (cut). Just below the presence of adipose cells (ad) with vacuoles in the cytoplasm (va). Scale bar = 100 µm. B. Detail of simple cubic epithelial cells (ep) with round-shaped nucleuses (n), where vacuoles can be seen (arrow), the cuticle a (cut) becomes thicker and the presence of a granular and acellular region can be observed (gr) in direct contact with the epithelium. Adipose cells (ad) are observed just under the epithelium, with vacuoles in the cytoplasm (va). Scale bar = 100 µm. C. Detail of the cuticle and (cut) and the thickening of the granular and acellular region (gr), which has structures in form of drops (arrow), in contact with the simple epithelium, now prismatic (ep), where vacuoles can be seen (arrow). Adipose cells (ad) are seen just under the epithelium, with vacuoles in the cytoplasm (va). Scale bar = 100 µm.

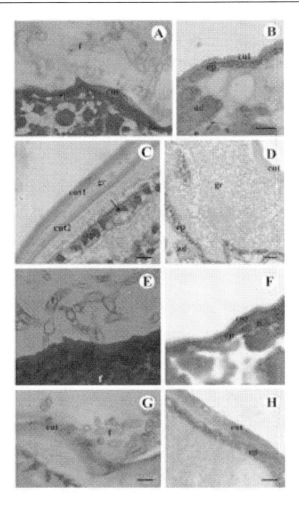

Figure 8. Histological sections of *Mycetarotes parallelus*, *Trachymyrmex fuscus* and *Myrmicocrypta* sp. ants represent successively more advanced degrees of separation of the cuticle during molting stained with Hematoxylin–Eosin. A. Histological sections of the simple scaly epithelium (ep) of *Mycetarotes parallelus* ant's larvae, where the cuticle can be seen (cut), as well as fungal hyphae (f), also emitting projections to the interior of the cuticle towards the epithelium. Scale bar = 100μm. B. Detail of simple cubic epithelium (ep) and of the cuticle (cut) of *Mycetarotes parallelus* ant's larvae. Scale bar = 100μm. C. Histological sections of simple cubic epithelium (ep) of *Trachymyrmex fuscus* ant's larvae, where vacuoles can be seen (arrow). It is possible to see the old cuticle (cut1), as well as the formation of the new cuticle (cut2) and between them the presence of a granular and acellular region (gr). Scale bar = 100μm. D. Detail of the granular and acellular region thickening (gr) located between the cuticle (cut) and the simple epithelium, now prismatic (ep) of *Trachymyrmex fuscus* ant's larvae. Adipose cells (ad) can also be seen. Scale bar = 40μm. E. Histological sections of *Trachymyrmex fuscus* ant's larvae, where the fungal hyphae can be seen (f), inclusive emitting projections to the interior of the cuticle towards the epithelium. Scale bar = 100μm. F. Detail of the simple cubic epithelium (ep) with oval nucleuses (n) and of the cuticle (cut) of *Trachymyrmex fuscus* ant's larvae. Scale bar = 100μm. G. Histological sections of *Myrmicocrypta* sp. Ant's larvae, where the cuticle can be seen (cut), as well as fungal hyphae (f), emitting projections to the interior of the cuticle towards the epithelium. Scale bar = 100μm. H. Detail of simple scaly epithelium (ep) and of the cuticle (cut) of *Myrmicocrypta* sp. ant's larvae. Scale bar = 100μm.

The application of morphological techniques revealed that *A. sexdens rubropilosa* (control), *M. parallelus*, *T. fuscus* and *Myrmicocrypta* sp. have the integument composed of: a) cuticle (outer), that in the present study was shown as a thin layer covering the individuals in their entirety, b) simple epithelium that varied from squamous to cubic, becoming prismatic in some cases, and, c) fat body cells supporting the previously described structures (Figures 7A, B, C; 9A-E).

The insect epidermal cells form an epithelium that covers their whole body and can present variations in shape according to the molt and intermolt periods and can change as they occupy different regions of the body (Sedlak and Gilbert, 1979).

The stages in the molt-intermolt process influence the morphophysiology of the larvae epithelial cells, which present a squamous shape during the intermolt process and become prismatic during the secretion and reabsorption when the molt occurs (Figures 7A, B, C; 9C, D). These cells also participate actively in the absorption of molt fluid via coated vesicles while they are secreting the new cuticle (Locke, 1985), which was observed in the present study in *A. sexdens rubropilosa* and *T. fuscus* larvae (Figures 7A, B, C; 8C). After the formation of the new cuticle and during the molt process the epithelial cells become squamous again, and this was clearly shown in *Myrmicocrypta* sp. larvae (Figure 8H). However, it was observed that turning to squamous did not exclude the cells activities, i.e., the new deposited cuticle would need maintenance to carry out its functions, e.g., protect the insect.

In the first stages of the insect molting process the substance initially released is called molt gel and does not present enzymatic activity, which will occur only after the deposition of the cuticuline layer (new cuticle) (Figures 7B, C; 8C, D; 9B). The exuvial space starts to contain a less viscous material due to the activation of this gel, now called molt fluid, which will act on the encoduticle, inner cuticle layer (Locke, 1982). At this stage the presence of several autophagic vacuoles can be observed in the interior of the cells and among epidermal cells, structures that would be involved in the recycling processes of cell organelles (Sedlak, 1982).

In *A. sexdens rubropilosa* larvae epithelial cells several vacuoles can be observed, indicating that these larvae will be in molt stage (Figures 7B, C). Several studies have demonstrated that the degradation of the old cuticle during the insect molting process occurs through the action of proteolytic enzymes released by the molt fluid. The proteases would remove the cuticular proteins exposing the cuticle to the action of chitinases (Dwivedi and Agrawal, 1995). In a more advanced molt stage it was observed that during the insect cuticle esclerotization process there would be the stabilization of the chitin matrix via pore channels and with this the deeper layers of the cuticle would not be affected (Hepburn, 1985).

Below the cuticular lamellae observed the presence of simple squamous epithelium of the epidermis of these larvae and the presence mitochondria of various shapes and sizes can be observed in the cytoplasm of the same (Figures 9C, D). The epithelial cells are very close to each other, and remain linked by cell junctions, especially those like interdigitations.

In the present study the fat body adipose cells were seen filling the subepidermal spaces of *A. sexdens rubropilosa* larvae (Figures 7A, B, C; 9D). In the intermolt stage these cells are larger due to the accumulation of reserve granules on the cytoplasm, probably of lipidic and proteic nature (Figure 7A). The presence of vacuoles can also be observed in the cytoplasm of these cells (Figures 7A; 9D). In the posterior stages where the larvae would probably be through the molt stage, the adipose cells would show cytoplasmatic structures that would suggest degeneration, since it was possible to observe the decrease of proteic and lipidic material in the form of granules concomitantly to the increase in the number of autophagic vacuoles (Figure 7C).

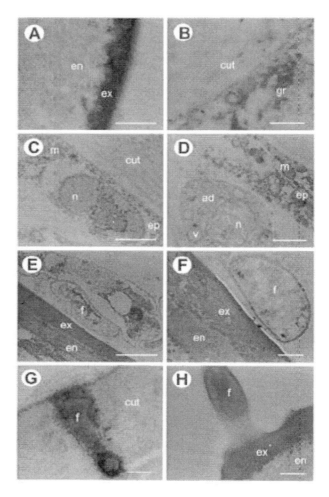

Figure 9. A. B. Electron micrograph of the larvae of *A. sexdens rubropilosa*. Bar = 16µm. C. D. Details of the squamous epithelium. Bar = 2µm. E. F. Larvae of *Mycetarotes parallellus*. E. Bar = 40 µm. F. Bar = 3 µm. G. Electron micrograph of the larvae of *Trachymyrmex fuscus*. Detail of the insertion of the fungi hyphae within the cuticle. Bar = 10µm. H. Electron micrograph of the larvae of *Myrmicocrypta* sp. Details of the projection of the fungi hyphae outside the cuticle. Bar = 10µm, cut = cuticle, en = endocuticle, ex = exocuticle, gr = granular region, ad = adipose cells, n = nucleus, v = vacuoles, f = fungi hyphae.

The present morphological study showed that in *M. parallelus*, *T. fuscus* and *Myrmecocrypta* sp. basal attine larvae the fungus that are deposited in the external side, i.e., on the surface of the larva, emit projections to the interior of the cuticle and of the epithelium, demonstrating an intimate fungi/larvae association (Figures 6D, E, F; 8A, E, G).

CONCLUSION

Literature has shown that researchers who study the behavior of ants, especially *Linepithema*, have tried to explain the social function of the dorsal protuberance in the immature of the genus. One of the first functions described is nutrition, once trophic interaction (by the queen or workers, representing a kind of nondestructive parental cannibalism) had already been seen in larvae and adults form several Ponerinomorfos group species. Although the larvae did not die, there was a rupture of their integument with the consequent release of hemolymph, which would be ingested by the adult. In other description, the larvae would present "hemolymph taps," specialized in conducting hemolymph to the surface of the larva for posterior consumption (Lacau, 2005; Masuko, 1986, 1989, 1990). Other studies about the behavior of Myrmicinae also showed intense interaction between adults and offspring through feeding retribution from larva to adult, which was named anal trophallaxis, i.e., once the larvae does not have a functional anus before reaching pupal stage, this kind of trophallaxis would function as a channel to release anal gland secretion, producing a liquid that would be nutritionally used by the workers (Hölldobler and Wilson, 1990). The morphological results presented in this study are actually the first information that can help to understand the morphological organization of *L. humile* larvae integument and indicate that the dorsal protuberance observed in the sixth abdominal segment apparently has no secretory function, as ducts or pore channels were not found, only small folds in the cuticle, which does not indicate a secretion release function. The secretion released by the flattened and the prismatic cells is composed of cuticle compounds and does not synthesize other kinds of compounds in social behavior. The epidermic cells from the dorsal protuberance would have their morphology modified from flattened to prismatic shape because of the need to produce a thicker cuticle layer, differently from the rest of the epidermis where the cuticle would be thinner, although the flat cells also take part in the production of secretion for the composition of the cuticle. Therefore, with the use of morphological techniques, the social function that can be attributed to the dorsal protuberance in *L. humile* larvae is the one of facilitator making it easier for the workers to transport the immatures through the different environments in the interior of the colony, a function that shall be proved in other studies that are already being carried out, analyzing the behavior of parental care as well as the workers' mandibles autonomy from the dorsal protuberance.

Regarding the presence of fungi covering the larvae of basal attine species, Ramos-Lacau et al. (2008) observed fungus in form of tufts in *C. transversus* species. These authors have registered the important role of the workers during larvae ecdyse, where they would lick the eggs and also immature ants (pupae and larvae) intensively which would allow the development of the fungus on the tegument, as the secretions could serve as culture medium for them. Soon after this procedure the workers would deposit fungus fragments from the main garden on the immature ant through the mandibles, antennas and front limbs.

Some authors have formulated hypotheses to explain the behavior of fungus deposition on the surface of basal attine larvae. It was suggested that the fungus could benefit the larvae protecting them against occasional pathogens and/or parasites by forming a physical or chemical barrier (Weber, 1972, Schultz and Meier, 1995). Other authors still argue that this fungus covering the larvae would facilitate the molt process, helping the old cuticle's degradation (Ramos-Lacau et al. 2008).

Although basal attine behavior of depositing fungus fragments on the larvae has been widely addressed in literature (Weber, 1972, Swartz, 1998, Schultz and Meier, 1995, Ramos-Lacau et al. 2008) there is no explanation as to how the relation fungus/larval cuticle would happen (either morphologically or physiologically), and it is unknown whether this association would be merely mechanical or if there would be any kind of physiological interaction between fungi and these ant larvae. Data obtained in this study through morphological techniques could therefore demonstrate that the fungus deposited on the surface of immature ant forms basal attine species maintain a close relationship to them, since the hyphae have the ability to disorganize the cuticular lamellas as well as being able to penetrate the interior of the insect through the emission of prolongations, transporting the cuticle and epithelium barriers, and exchanging substances.

REFERENCES

Camargo, RS; Lopes, JFS; Forti, LC. (2006). Behavioural responses of workers towards worker-produced male larvae and queen-produced worker larvae in *Acromyrmex subterraneus* brunneus Forel, 1911 (Hym., Formicidae). *J. Appl. Entomol.*, 130, 56-60.

Currie, CR. et al. (2003). Ancient tripartite coevolution in the attine ant-microbe symbiosis. *Science*, 299, 386-388.

Giraud, T; Pedersen, JS; Keller, L. (2002). Evolution of supercolonies: The Argentine ants of southern Europe. *Proc. Natl. Acad. Sci. USA*, 99, 6075–6079.

Hackman, RH. (1964). Chemistry of the insect cuticle. In: Rockstein, M. (Ed.), *Physiology of Insecta*. New York, Academic Press, pp. 471-506.

Hackman, RH. (1971). *The integument of arthropoda*. In: Florkin M., Sheer, B.T. (Eds.), Chemical Zoology New York, Academic Press, pp. 1-62.

Hölldobler, B; Wilson EO. (1990). *The ants*. Cambridge: The Belknap Press of Harvard University, pp. 732.

Junqueira, LC; Carneiro, J. (2004). *Histologia básica*. Rio de Janeiro: Guanabara Koogan. pp. 540.

Lacau, S. (2005). *Morphologie et Systématique du genre Typhlomyrmex Mayr, 1862 (Formicidae: Ectatomminae)*. Dissertation. Paris: Muséum National d'Histoire Naturelle.

Lattke, JE. (1997). Revisión del género *Apterostigma* Mayr (Hymenoptera: Formicidae). *Arq. Zool.*, 34, 121-221.

Locke, M. (1985). The structural analysis of post-embryonic development, in: kerkut, G.A., Gilbert, L.I. (Ed.), *Comprehensive insect physiology, biochemistry and pharmacology*. Oxford, Pergamon Press, pp. 87-149.

Lopez, JFS, et al. (2005). Larval isolation and brood care in *Acromyrmex* leaf-cutting. *Insectes Sociaux*, 52, 333-338.

Masuko, K. (1986). Larval hemolymph feeding: A non-destructive parental cannibalism in the primitive ant *Amblyopone silvestrii* (Wheeler) (Hymenoptera: Formicidae). *Behav. Ecol. Sociobiol.*, 19, 249–255.

Masuko, K. (1989). Larval hemolymph feeding in the ant *Leptanilla japonica* by use of a specialized duct organ, the larval hemolymph tap (Hymenoptera: Formicidae). *Behav. Ecol. Sociobiol.*, 24, 127–132.

Masuko, K. (1990). Behaviour and ecology of the enigmatic ant *Leptanilla japonica* Baroni Urbani (Hymenoptera: Formicidae: Leptanillinae). *Insectes Sociaux*, 37, 31–57.

Mueller, UG; Rehner, SA; Schultz, TD. (1998). The evolution of agriculture in ants. *Science*, 281, 2034-2038.

Murakami, T; Higashi, S. 1997. Social organization in two primitive attine ants, *Cyphomyrmex rimosus* and *Myrmicocrypta ednaella*, with reference to their fungus substrates and food sources. *J. Ethol.*, 15, 17-25.

Pedersen, JS; Krieger, MJB; Vogel, V; Giraud, T; Keller, L. (2006). Native supercolonies of unrelated individuals in the invasive Argentine ant. *Evolution*, 60, 782–791.

Queller, DC. (2000). Pax Argentinica. *Nature*, 405, 519–520.

Ramos-Lacau, LSR, et al. (2008). Morphology of the eggs and larvae of *Cyphomyrmex transversus* Emery (Formicidae: Myrmicinae: Attini) and note on the relationship with its symbiotic fungus. *Zootaxa*, 23, 37-54.

Reynolds, SE; Samuels, RI. (1996). Physiology and biochemistry of insect molting fluid. *Adv. Insect. Physiol.*, 26, 157-232.

Richards, AG. (1951). *The Integument of Arthropods*. Minneapolis, University of Minnesota Press, pp. 400.

Rossini, AS. (1997). Alteracões morfológicas dos componentes tegumentares de estágios imaturos de *Melipona quadrifasciata anthidioides* LEP (Hymenoptera, Apidae). 118f. Tese (Doutorado em Zoologia). Rio Claro: Universidade Estadual Paulista.

Schultz, TR; Meier, RA. (1995). Phylogenetic analysis of the fungus-growing ants (Hymenoptera: Formicidae: Attini) based on morphological characters of the larvae. *Systematic Entomology*, 20, 337-370.

Shattuck, SO. (1992). Generic revision of the ant subfamily Dolichoderinae (Hymenoptera, Formicidae). *Sociobiology*, 21, 1–181.

Smith, D. (1968). Insects cells: their structure and functions. Oliver and Boyd, Edinburgh, pp. 372.

Solis, DR; Fox, EGP; Rossi, ML; Bueno, OC. (2010). Description of the immatures of *Linepithema humile* Mayr (Hymenoptera: Formicidae). *Biol. Res.*, 43, 19-30.

Tsutsui, ND; Suarez, AV. (2003). The colony structure and population biology of invasive ants. *Conserv. Biol.*, 17, 48–58.

Tsutsui, ND; Suarez, AV; Holway, DA; Case, TJ. (2000). Reduced genetic variation and the success of an invasive species. *Proc. Natl. Acad. Sci. USA*, 97, 5948–5953.

Villesen, P; et al. (2004). Evolution of ant-cultivar specialization and cultivar switching in *Apterostigma* fungus-growing ants. *Evolution*, 58, 2252-2265.

Wang, Y; Mueller, UG; Clardy, JC. (1999). Antifungal diketopiperazines from the symbiotic fungus of the fungus-growing ant *Cyphomyrmex minutus*. *J. Chem. Ecol.*, 25, 935-941.

Weber, NA. (1972). *Gardening ants: The Attines*. Philadelphia, the American Philosophical Society, pp. 146.

Wheeler, WM. (1907). The fungus-growing ants of North America. *Bull. Am. Mus. Nat. Hist.*, 23, 669-807.

Wigglesworth, VB. (1948). The determination of characters at metamorphosis in *Rhodnius prolixus* (Hemiptera). *J. Exp. Biol.*, 25, 1-24.

Wigglesworth, VB. (1957). The action of growth hormones in insects. *Symp. Soc. Exp. Biol.*, 11, 204-227.

Chapter 8

Biology and Life Cycle of Thitarodes Pui (Lepidoptera, Hepialidae), a Host of the Caterpillar Fungus Ophiocordyceps Sinensis in the Tibetan Plateau

*Zhi-Wen Zou[1,2], Jun-Feng Li[2] and Gu-Ren Zhang[2] **

[1]School of Life Sciences and Food Engineering, Nanchang University,
Nanchang, China
[2]State Key Laboratory for Biocontrol and Institute of Entomology,
Sun Yat-Sen University, Guangzhou, China

Abstract

Larvae in the genus *Thitarodes* (Lepidoptera: Hepialidae) are host to *Ophiocordyceps sinensis,* a unique entomophagous fungal parasite that occurs principally in the alpine meadow environment of the Tibetan Plateau and for centuries a widely known and valuable invigorant in traditional Chinese medicine. *T. pui* is mainly distributed at an altitude of 4100 ~ 4650 m in the alpine meadows and alpine shrub meadows of Mt. Segyi La in Tibet. Larvae feed on plants herbaceous roots and humus fragments. Larval development lasts three to four years (about 1095 ~ 1460 d), including 41 ~ 47 d for the egg, 990 ~ 1350 d for the larva, 35 ~ 41 d for the pupa, and 3 ~ 8 d for the adult. Larval growth involves seven to nine instars. Larval head capsule width is the principal parameter to distinguish each larval instar. The ratio of head capsule width to body length and weight decreased with increasing age. The 7th instar pupates into male adults, the 9th instar linto female adults, and the 8th instar larvae into both males and females. Larvae showed an aggregated distribution where larval densities had their greatest concentration between altitudes of 4100 ~ 4650 m. Pupation occurred in the end of April to early May. Adults emerged during the day at any time between sunrise and sunset from late June to mid July. Courtship occurred between 21:00 to 21:30 h followed by mating last from about one to several hours. The ratio of males to females was 1.5:1. The average oviposition was 768 ± 206 eggs per female. The life table of an experimental

---

* Author for correspondence: Email: zhanggr@mail.sysu.edu.cn.

T. pui population was constructed through laboratory rearing. The total survival rate of the experimental population was 2.6%, and the population trend index was 7.95 which indicated that the population size of the next generation will be 7.95 times greater.

Keywords: *Thitarodes pui*, biology, life cycle

INTRODUCTION

Ophiocordyceps sinensis (Berk.) G.H. Sung, J.M. Sung, Hywel-Jones and Spatafora (Sung *et al.*, 2007) is a well known traditional Chinese medicine and one of the economic resources for farmers and herdsmen who collect *O. sinensis* in the paramos of the eastern and central Tibetan Plateau. In recent years, declines in the resource along with increasing market demand and soaring prices has roused widespread concern about future sustainable utilization. As the host of *O. sinensis*, it is important to investigate the species and distribution of genetic qualities of *Thitarodes* which larvae provide the nutrient source and trophic foundation of the fungus. Understanding the biology and ecology of *Thitarodes* is the key to the problem of solving the sustainable utilization of *O. sinensis* as an economic resource.

In order to better understand and sustainably utilize this specialized natural medicine resource on the Tibetan Plateau, our research group set up an experimental station in 2006 at 4,156 m on Mt. Segyi La (29°37′N, 94°37′E) in the Nyainqentanglha Range. *T. pui* (Zhang *et al.*, 2007) (Lepidoptera: Hepialidae) represented a newly discovered species near the station firstly reported in China as *Hepialus pui* (Zhang et al., 2007) and later transferred to genus *Thitarodes* (Zou, 2009; Zou et al., 2010).

Figure 1. *T. pui* larva in the soil.

It is mainly distributed at altitudes of 4100 ~ 4650 m in the alpine meadows and alpine shrub meadows. The larvae live in the soil (Figure 1) and feed on non-woody roots and soil humus. In the three year life cycle, the pupa, adult and egg stage completed in four months from May to August of the same year. Larvae overwinter in their soil tunnels for three or four winter seasons in the frozen soil. In this research, additional biological characteristics and aspects of the life cycle of *T. pui* were studied in order to provide information for the sustainable harvesting of *O. sinensis* on Mt. Segyi La.

1. MATERIALS AND METHODS

1.1. Habitats of *T. Pui*

Mt. Segyi La is located in the east of Linzhi County in Tibet, and belongs to Nyainqentanglha Range. It is the watershed of Niyang River and Polong Zangbo River. The annual rainfall is more than 650 mm.

T. pui is distributed at altitudes of 4100 to 4650 m in the alpine meadows and alpine shrub meadows around the station in Mt. Segyi La (Figure 2).

Figure 2. Habitats of *T. pui.*

The principal vegetation comprises low shrubs and herbs, and the soil surface is covered with a dense mat of plant roots and rich in associated organic matter. There are abundance rain and moist soil in summer, while it is cold and frozen soil can be formed in winter. The temperature difference between day and night is very large throughout the year.

1.2. Materials

Most specimens, including paired adults and most of the pupae and larvae, were collected from Mt. Segyi La. Other pupae and larvae were raised on a diet of carrots in the insectary of the station.

1.3. Methods

1.3.1. Sampling

T. pui larvae were randomly collected from the field in each month from a 1m × 1m sampling plot. In order to protect the habitat while sampling, the grass layer was unraveled during sampling. Soil was then excavated to a depth of 50 cm, all larvae were collected, and the soil was replaced and covered over again by the grass layer. Larvae were identified by measuring head capsule width. The survey in each sample plot was repeated three times and over a three year period.

1.3.2. Eggs

Mating pairs were collected and each placed into a silk yarn bag (70 cm × 34 cm × 34 cm). Eggs deposited by the moths were collected immediately and color changes were observed after oviposition. In order to detect environmental effects on hatching rate, eggs from the same moth were divided into two groups and placed into the insectary (dark, room temperature, RH 60%- 80%) or in the field (shade, no rain). Eggs of each group were placed on three surface materials: double filter paper, humus, and paper cup (control group). Each treatment was repeated three time and 50 eggs were used for each replicate. The hatching of eggs was observed two times each day.

1.3.3. Larvae

1.3.3.1. Life Cycle and Biology of Larvae

Larval biology was investigated by field sampling and artificial rearing in insectary. From April to October in each year, larvae were surveyed by random sampling in the habitats for different altitudes, vegetation types and slope. Larval station in the soil, population density, head capsule width, and body length and width of larvae were recorded.

Larvae collected from the field were individually reared over a three year period on carrot root in a small plastic cup, about 50 individuals for each instar. Survival and molting rates were regularly checked. The head capsule width, length and weight of larvae are measured when larval head color turned brown after molting.

1.3.3.2. Spatial Distribution of Larvae

The spatial distribution of larvae in various habitats was investigated from June to August each year. Seven sampling plots representing different habitats were selected as follows: valley (4184±4m, 29°36.301′N, 94°36.547′E),lakeside (4620±7m, 29°38.905′N, 94°35.304 ′E), slope beside the big turning of the national road 318 (4369 ±13m, 29°38. 190′N, 94°37.383′E), enclosure in sunny slope (4144±13m, 29°35.687′N, 94° 35.981′E), 113 classes (4205 ± 5m, 29°36.329′N, 94°36.261′E), sunny slope (4190±13m, 29°36.455′N, 94°36.526′E), shady slope (4200±13 m, 29°35.869′N, 94°36.275′E).

1.3.3.3. Pupae

Sixty pre-pupae collected from the field and the insectary were each placed into a 6 cm diameter and 7cm high glass beaker and surrounded by damp humus, which was put into a plastic bucket (25cm diameter, 23cm high) containing moist moss, with dark and RH 80%. Pupation was observed and pupae stage was recorded.

1.3.3.4. Adults

Mating behavior was observed in the field and the insectary. More than one hundred individual males and females and thirty mating pairs were placed inside a mosquito net cage (195 cm × 114 cm × 193 cm). The oviposition behavior of female, number of laid eggs and the longevity of each adult was recorded. After females died they were dissected to record the number of remaining eggs.

1.3.4. Life Table of Experimental Population

Life history data for each developmental stage was recorded and a life table was constructed for the experimental population. Larval development was divided into four phases: 1st-3rd, 4th-5th, 6th-7th and 8th-9th, each phase lasting about one year.

1.4. Data Analysis

1.4.1. Spatial Distribution of Larvae

Spatial distribution of larvae was analyzed by the following four aggregation indices:

$$I = \frac{S^2}{\bar{x}} - 1$$

(1) Index of clumping of David and Moore (1954), where $I = \frac{S^2}{\bar{x}} - 1$

If I=0, <0 and >0, the distribution is random, uniform, or clumped, respectively.
(2) Cassie's index (Cassie, 1962), where C=1/K
If C=1, <1 and >1, the distribution is random, uniform, or clumped, respectively.

$$= \frac{\sum_{i=1}^{N} x_i(x_i-1)}{\sum_{i=1}^{N} x_i}$$

(3)Mean crowding (Lioyd, 1967), where $M^*/m = \frac{\sum_{i=1}^{N} x_i(x_i-1)}{\sum_{i=1}^{N} x_i} / m$ $M*/m$ /m. x_i is the

number of individuals in the plot *i*.

If $M*/m$=1, <1 and >1, the distribution is random, uniform, or clumped, respectively.

$$I_\delta = \frac{\sum_{i=1}^{Q} n_i(n_i - 1) \cdot Q}{N(N-1)}$$

(4) Index of dispersion (Morisita, 1959), where $I_\delta = \frac{\sum_{i=1}^{Q} n_i(n_i-1) \cdot Q}{N(N-1)}$ .Q is the number of samples, n_i is the number of individual in the plot *i*, N is the total number of individuals.

If I_δ=1, <1 and >1, the distribution is random, uniform, or clumped, respectively.

1.4.2. Life Table of Experimental Population

Total survival rate of generation= $S_1 \times S_2 \times S_3 \times S_4 \times S_5 \times S_6$ (a)

Where S_1-S_6 is the mean survival rate of eggs, 1st-3rd, 4th-5th, 6th-7th, 8th-9th instar and pupae, respectively.

The population trend index was calculated by Morris-Watt mathematical model:

$$I= S_1 \times S_2 \times \ldots S_n \times F \times P_F \times P_\female$$

In this equation, S_1-S_n is the survival rate of each stage, F is the standard fecundity, P_F is the percentage of females which reached the standard, $P_\female$ is the rate of females. The population trend index was calculated by the modified formula:

$$I= S_1 \times S_2 \times \times S_3 \times S_4 \times S_5 \times S_6 \times P_\female \times N_E$$ (b)

In this equation, S_1-S_6 mean the survival rate of eggs, 1st-3rd, 4th-5th, 6th-7th, 8th-9th instar and pupae, respectively. N_E is the average fecundity.

1.4.3. Analysis of Variance on Egg Hatch Rate

Hatching rate in three substrates of two habitats was analyzed with single factor analysis of variance by software Statistica (version 9.0, Statsoft, USA). Significant difference tests were analyzed by the least significant difference (LSD) method.

2. RESULTS AND ANALYSES

2.1. Life Cycle

Larval development lasted three to four years (about 1095 to 1460 d), including 41- 47 d for the egg, 990- 1350 d for the larva, 35- 41 d for the pupa, and 3- 8 d for the adult. Pupation occurred from the end of April to early May.

Table 1. Distribution of developmental stages of *T. pui* in a year

Month	Jan to Apr			May			Jun			Jul			Aug			Sep to Dec		
Mean	- 4.66°C			5.58°C			7.83°C			9.32°C			6.12°C			- 0.46°C		
a period of	1st	2^n	3^r	1st	2^n	3^r	1st	2^n	3^r	1st	2^n	3^r	1st	2^n	3^r	1st	2^n	3^r
egg										·	·	·	·	·	·			
larva	—	—	—	—	—	—	—	—	—	—	—	—	—	—	—	—	—	—
pupae				△	△	△	△	△	△	△	△							
adult										+	+	+						

".”--egg,” “—”--larva, “△”--pupae, “+” ----adult.

The exact dates varied slightly each year due to changes in seasonal weather conditions. Adults emerged during the day at any time between sunrise and sunset, from late June to mid July. Adult longevity in the laboratory was 3- 8 d. Distribution of developmental stages of *T. pui* in a year was shown in Table 1.

2.2. Biology of Eggs

Eggs are slightly ellipse in shape, 0.65 - 0.83 mm long and 0.42 - 0.63 wide. The egg color started to change from the creamy-white color at 0 h to glossy-black at 4- 6h, and entirely black at 10- 14 h after oviposition.. Egg stage lasted 41- 47 d, affected by the ambient temperature (Table 2). Hatch rates were significantly higher in the lab than in the wild. The low hatching rate in the wild might result from the drastic changes in temperature and humidity.

Table 2. Effects of environmental condition on the hatching rate of *T. pui* eggs

Environment	Temperatures(°C)	RH (%)	Background	Num of eggs	Hatching rate (%)	Duration (days)
Insectary	10~17	65~75	Filter paper	50×3	92.0±2.0a	41~45
			Humus soil	50×3	86.0±1.7b	41~47
			Paper cup	50×3	94.0±1.0a	42~45
Field	4~20	41~78	Filter paper	50×3	30.0±1.7A	41~45
			Humus soil	50×3	20.0±2.0B	41~46
			Paper cup	50×3	28.0±2.6A	42~45

Notes: The same small or capital letter following the mean of hatching rate in the same environmental treatment means no significant difference between replicates.

2.3. Biology of Larvae

2.3.1. Larval Development

There were 7- 9 instars during larvae development and there were overlapping generations present in any one sample (Figure 3). Pupation occurred in the 7th to 9th instars with 7th instar larvae pupating into male adults, 9th instar larvae into female adults, and the 8th instar larvae into male or females. Larvae fed more frequently and grew faster during April to October when they moult two to three times.

Larval activity at the beginning of October gradually reduced until inactive by the end of October. All larvae after the 3rd instar could overwinter. Each stage of larvae changed greatly if they were happened to overwinter. Larvae recommenced feeding in the following April when the soil began to thaw.

The instar was able to be distinguished by head capsule width whereas variation in body length and weight overlapped between instars.. The increasing ratio of larval head capsule width, body length and weight ratio of larvae decreased with increasing age (Table 3- 5).

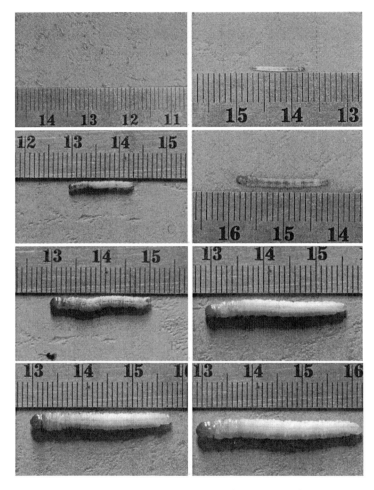

Figure 3. Different instars of *T. pui* larvae A. 1st instar; B. 2nd instar; C. 3rd instar; D. 4th instar; E. 5th instar; F. 6th instar; G. 7th instar; H. 8th instar; I. 9th instar; J. prepupa.

Table 3. Head capsule width of *T. pui* larvae\

Instar	Number	Range (mm)	Mean±SD(mm)	Ratio of increase
1st	30	0.35~0.60	0.45±0.14	
2nd	30	0.80~1.30	1.10±0.13	2.44
3rd	40	1.44~1.80	1.61±0.10	1.46
4th	35	1.94~2.38	2.17±0.11	1.34
5th	31	2.42~2.90	2.71±0.17	1.25
6th	36	2.92~3.26	3.11±0.10	1.15
7th	49	3.28~3.64	3.46±0.11	1.11
8th	43	3.58~3.96	3.76±0.10	1.09
9th	33	4.00~4.42	4.19±0.13	1.11
Average				1.37

Table 4. Body length of *T. pui* larvae

Instar	Number	Range (mm)	Mean±SD(mm)	Ratio of increase
1st	30	0.40~0.70	0.60±0.16	
2nd	30	1.30~1.70	1.40±0.13	2.33
3rd	40	1.40~2.30	1.70±0.19	1.21
4th	35	1.80~2.70	2.30±0.21	1.35
5th	31	2.40~3.40	2.90±0.33	1.26
6th	36	2.50~4.00	3.20±0.33	1.10
7th	49	2.40~4.30	3.40±0.41	1.06
8th	43	3.30~4.70	4.00±0.37	1.18
9th	33	4.10~5.10	4.60±0.24	1.15
Average				1.33

Table 5. Individual weight of *T. pui* larvae

Instar	Number	Range (mm)	Mean±SD(mm)	Ratio of increase
2nd	30	0.01~0.02	0.01±0.00	
3rd	40	0.01~0.05	0.03±0.01	1.93
4th	35	0.04~0.11	0.07±0.02	2.69
5th	31	0.06~0.20	0.13±0.04	1.89
6th	36	0.08~0.31	0.19±0.05	1.45
7th	49	0.13~0.41	0.24±0.05	1.23
8th	43	0.23~0.58	0.36±0.08	1.50
9th	33	0.41~0.58	0.50±0.05	1.38
Average				1.72

2.3.2. Distribution of Larvae

T. pui is distributed at altitudes of 4100 to 4650 m in the alpine meadows and alpine shrub meadows of Mt. Segyi La in Tibet in soils with a pH of 6.2- 6.8. Larval densities were higher at lower altitudes, and decreased with increasing altitude. Field surveys showed that larvae of *T. pui* are distributed in aggregated distribution pattern in different habitats (Table 6).

Table 6. Aggregation index of *T. pui* larvae

Sampling spot*	Mean (m)	S^2	K	C	I	M*/m — M*	M*/m — M*/m	I_δ
1	5.33	85.08	0.06	2.81	14.98	20.17	3.79	3.79
2	6.48	114.20	0.06	2.57	16.64	22.97	3.55	3.55
3	1.19	3.52	0.34	1.65	1.96	3.08	2.59	2.64
4	5.80	47.43	0.12	1.24	7.18	12.81	2.21	2.22
5	2.20	12.16	0.18	2.07	4.54	6.60	3.01	3.04
6	8.71	137.04	0.06	1.69	14.73	23.29	2.67	2.68
7	29.22	4283.03	0.01	4.98	145.58	171.87	5.88	5.89

* Sampling spot: 1. Sunny slope; 2. Shady slope; 3. Slope beside the big turning of the national road 318 (Turning); 4. Lakeside; 5. Back of maintenance gang house 113; 6. Valley; 7. Enclosure in sunny slope.

2.3.3. Larval Habit

Newly hatched larvae is less than 2 mm in length. The first instar larvae move about as soon as leaving egg shells and looked for a suitable habitat in soil surface. Each larva lives in a vertical or slightly inclined soil tunnel. The diameter and depth of tunnels are related with larval instars. The inner surface of the tunnel was covered with a layer of silk that is coated with soil particles on the external surface. The end of the tunnels are is slightly enlarged where larvae can turn around within. Collembola are also usually found within the tunnels.

The body of 1st and 2nd instar larvae is nearly transparent and the larvae could be reared in together. Cannibalism may occur after the 3rd instar. Investigation showed that larvae are polyphagous and feed on not only herbaceous roots but also soil humus with molecular analysis of food residues in the foregut.

2.4. Biology of Pupae

Pupation usually began in May, but a few could begin at the end of April. The pupal stage lasted 37.5 ± 1.9 d (n = 40).

Pupation goes through a pre-pupal stage with thicker and shorter bodies in soil tunnels. The pupa is initially white, sequentially turns yellow, brown, and black. In soil tunnels, pupae move up and down in response to changes in temperature and humidity. Pupae move closer to the ground surface when emergence was coming.

Pupae are extremely sensitive to humidity changes, and the optimal humidity is about 80% for pupal development. Pupae are prone to death for lack of moisture caused by dramatic changes in humidity

2.5. Biology of Adults

2.5.1. Emergence

Most adults emerge from late June to mid July, but there are some differences between years and among different climatic conditions.

Figure 4. The emergence of *T. pui* adults. A. Adult out of puparium B. Adult spreading wings.

Temperature is the key factor influencing the emergence date. Generally, emergence occurres earlier in the insectary than in field, and pupae in slopes with greater exposure to sun emerge earlier than those in shade. Emergence may occur throughout the whole day, but peak emergence appeare at about 17:00 h. It is about a minute for adult to emerge from pupa. The body surface of the fresh adult is damp and the partially expanded wings are close to the body.

The moth would climb up a nearby plant where wing expansion will take place for about 10- 20 min later (Figure 4), and then hides in a place until nightfall. The sex ratio of male to female is about 1.5:1.

2.5.2. Courtship and Mating

Females climbe to higher plant leaf and release sex pheromone by shaking of wings (Figure 5) to attract males at 21:00 to 21:30 in evening. And then, males begin to fly and search for female by flying. There are often several males surrounding one female.

Figure5. Mating of T. pui adults. A. female vibrating wings; B. mating adults.

Only one male can get the opportunity to mate with the female. As soon as mating occurres, the female stops vibrating of wings and the frustrated males fly away. Courtship lasts no more than 30 min. Mating lasts from 1 to 7 h with an average of 2.5 h. Adults usually mate only one times besides twice in rare females (less than 2%).

2.5.3. Oviposition

Females normally begin to lay eggs immediately after mating. They don't select their oviposition sites, just deposite their eggs on the ground by swinging their abdomen. The average oviposition is 768±206 eggs (n = 35), with 50% of total eggs being laid in the first ovipositon. 98% of eggs is eventually laid. Rare eggs, about 15±14 (n= 35) still stay in the dead females. Virgin females also lay about 90% of their eggs.

2.5.4. Longevity

Average longevity of females was 5.5±1.1d, males 5.2 ± 1.3 d (Table 7)

Table 7. Longevity of *T. pui* adults

Longevity (d)	3	4	5	6	7	8	Mean	SD
Males (n=30)	2	8	7	8	4	1	5.2	1.3
Females (n=42)	3	2	16	15	5	1	5.5	1.1

2.6. Life Table of Experimental Population

The life table of experimental population of *T. pui* is shown in Table 8.

Table 8. The life table of *T. pui* experimental population

Stadium (x)	Original individuals (N)	Survival number (L_x)	Survival rate (S_x)	Death number (d_x)	Mortality (q_x)
egg	450	408	90.7%	42	9.3%
1st-3rd instar	60	26	43.3%	34	56.7%
4th-5th instar	54	21	38.9%	33	61.1%
6th-7th instar	63	32	50.8%	31	49.2%
8th-9th instar	71	27	38.0%	44	62.0%
pupae	65	57	87.7%	8	12.3%
adult	female: male		Rate of female	Average fecundity	
	1:1.5		40%	768±206	

The total survival rate of the experimental puoplation was only 2.6%, and the population trend index (I) was 7.95, suggesting that the next generation individuals would be 7.95 times the tested generation number although the total survival rate was so low.

CONCLUSION AND DISCUSSION

T. pui is endemic to Mt. Segyi La where it takes three to four years to complete a generation which is similar to other *Thitarodes* species in China (Wang et al., 2001; Yin et al., 2004). *T. pui* larval development with seven to nine instars was different from other *Thitarodes* species larvae with about six instars (Yang et al., 1996; Wang et al., 2001; Yin et al., 2004). Especially, the 7th instar larvae pupated into male adults, the 9th instar larvae into female adults, and the 8th instar larvae into males or females. There are no references described the relationship between their instars and sex. It might be related with their sex determination and more nutrients for female development.

Biological characteristics of *T. pui* such as the diet and spatial distribution of larvae, adult mating and oviposition are similar to the other host species of *O. sinensis*, but differed in the peak of eclosion, courtship and mating times, sex ratio, and fecundity (Yin et al, 2004; Wang et al, 1995, 2001; Huang et al, 1992; Yang, et al., 1991; Zhang, 1988; Zhao et al, 1989;). The sex rate of male to female in *T. pui* is higher than that (1:1) in other *Thitarodes* species in China (Liu et al., 2005). Why males are much more than females is an interesting question worthy of further investigation.

The stable environment maintained in the insectary significantly increased the egg hatching rate. This approach for egg rearing followed by field release of first instar larvae might be possible to increase the larval density in field, and this in turn could result in more hosts being available for *O. sinensis*. In addition, ambient temperatures significantly affected larval development of *Thitarodes* (Yang et al., 1996), specially overwintering larvae were frozen for at least six months. If rearing in artificial conditions, the frozen time could be greatly reduced, and therefore possible to accelerate larval development and survival.

Most published records indicated that the herbaceous roots of plants adjacent to the larval tunnel were the principal food for *T. pui* larvae (Yang et al., 1996; Chu et al., 2004; Liu et al., 2006). This is the general pattern for Hepialidae where larvae either bore within the host plant (such as rhizomes, roots and stems), feed on roots that are exposed by the tunnel, or feed on surface vegetation where the tunnel is open to the ground surface and covered over by a canopy of silk and debris (McCabe and Wagner, 1989; Grehan, 1989, Tobi et al., 1989). As in many other Hepialidae, the larvae of *T. pui* are polyphagous on the herbaceous roots of a large number of plant species. The vegetation compositions in sample areas were investigated and the ITS were cloned and sequenced from 26 species of constructive plants. The results showed that *T. pui* larvae fed on the herbaceous roots of almost all alpine plants representing 16 families and 24 genera. The preferred hosts were *Ranunculus brotherusii*, *Cyananthus macrocalyx*, *Juncus leucanthus* and *Veronica ciliate* (Lei et al., 2011). Carrot is the ideal substitute for rearing larvae in the insectary as this plant will allow larvae to successfully complete development (Li et al., 2011). New evidence from the stable carbon isotope data of *T. pui* larvae and its closely substance showed that larvae feed on not only tender root but the soil humus fractions (Chen et al., 2009). Previously it was thought that tender plant roots provided the only food source in the field. The use of humus might contribute to the production of inexpensive high-quality food for large-scale artificial rearing of larvae.

It is difficult to accurately identify instars for *T. pui* larvae live in the field. From field survey and insectary rearing, head capsule width follows Dyar's law which is provides a correlated parameter for each stadium. But increasing rate of the head capsule width decreased with stadium increasing, the average increase rratio was 1.22, less than 1.4 in Dyar's law, which may be related to the environmental conditions. The survey on the head capsule of *O. sinensis* showed which larval stage was most susceptive for infecting of *O. sinensis*. It is important to find optimum instar for artificial producing of *O. sinensis*.

The rate of total larval survival in the insectary was very low, only 2.6%, but the survival rate in the field might be even lower where it could also be impacted by parasitism and disease. The population trend index (I) of the experimental population was 7.95, which indicated that the *T. pui* population increased 7.95 times even with the very low survival rate because of the high fecundity which averaged 768±206 eggs per female. Larval mortality was higher than in the other stages due to infection by pathogenic microorganisms resulting in over 80% mortality. Antibiosis and sterilization is therefore the key to successful larval rearing. The extended development period of *T. pui* combined with the limited experimental time limited the ability to develop a comprehensive life table at this time. The role of environmental factors affecting population fluctuations was also beyond the resources of the present study and also needs further investigation. Habitats of *T. pui* are the alpine meadows and alpine shrub meadows above an altitude of 4100 m. The habitats are complex with variations in temperature, slope, and aspect, with geographical isolation and frequent inclement climatic conditions (Yang et al., 1996; Zhang et al., 2007; Li *et al.*, 2011). The increased body weight due to the large number of eggs and the short adult life span limited the potential flight range of *Thitarodes* females. If this low dispersability of females along with topographic and climatic variability of the Tibetan Plateau has resulted in extensive genetic isolation, it is possible that many more species remain to be discovered as indicated by the present evidence for different *Thitarodes* species on different mountains, and even between different slope aspects and altitudes within the same mountain (Yang *et al.*, 1996; Liu *et al.*, 2005). This species diversity may provide an abundant resource for the sustainable utilization of *O. sinensis* (Gu et al., 2006).

ACKNOWLEDGMENTS

This research was supported by the National Key Technology R and D Program (2011BAI13B06, 2007BAI32B05, 2007BAI32B06) and the National Natural Science Foundation of China (31160081). We want to express our gratitude to Dr. John R. Grehan, at the Buffalo Museum of Science for his constructive criticism on the manuscript.

REFERENCES

Chen D, Yuan J P, Xu S P, et al., 2009. Stable carbon isotope evidence for tracing the diet of the host *Hepialus* larva of *Cordyceps sinensis* in the Tibetan Plateau. *Science in China (Series D: Earth Sciences,* 52(5):655-659.

Chu H F, Wang L Y, Han H X, 2004. Fauna Sinica, Insecta, Vol.38, Lepidoptera, Hepialidae and Epiplemidae. Science Press, Beijing. 1-194.

Grehan J R, 1989. Larval feeding habits of the Hepialidae (Lepidoptera). *Journal of Natural History*, 23:803-824.

Gu D X, Zhang G R, Wang J H, et al., 2006 A review and prospect on the studies of *Cordyceps sinensis* (Berk) Sacc. J*ournal of Chinese Institute of Food Science and Technology*, 6(2): 141-145.

Huang T F, Fu S Q, Luo Q M, 1992. Bionomics of *Hepialus gonggaensis* from Kangding. *Acta Entomologica Sinica*, 35(2): 250-253.

Lei W, Chen H, Zhang G R, et al., 2011. Molecular identification and food source inference of constructive plants, native to the *Ophiocordyceps sinensis* habitat. *African Journal of Biotechnology*, 10 (2):159-167.

Li J F, Zou Z W, Liu X, et al., 2011. Biology of *Thitarodes pui* (Lepidoptera, Hepialidae), a host species of *Ophiocordyceps sinensis* . *Journal of Environmental Entomology*, 33(2): 195- 202.

Liu F, Wu X L, Yin D H, *et al.*, 2006. Review on the species and their distribution of the genus *Hepialus*, *Chinese Traditional Medicine Research of Chongqing*, (1): 47- 50.

McCabe T L, Wagner D L., 1989. The biology of *Sthenopis auratus* (Grote) (Lepidoptera: Hepialidae). *Journal of the New York Entomological Society*, 97:1-10.

Sung G H, Hywel-Jones N L, Sung J M, et al., 2007. Phylogenetic classification of *Cordyceps* and the clavicipitaceous fungi. *Studies in Mycology*, 57(1): 5–59.

Tobi D R, Wallner W E, Parker B L., 1989. The conifer swift moth, *Hepialus gracilis*, and spruce-fir decline. In: Proc. US/FGR Research Symposiun: Effects of atmospheric pollutants on the spruce-fir forests of the eastern United States and Federal Republic of

Germany. *United States Forest Service North East Forest Experiment Station., General Technical Report*, NE-120, 351-353.

Tu Y Q, Ma K S, Zhang D L, 2009. A new species of the Genus *Hepialus* (Lepidoptera: Hepialidae) from China. *Entomotaxonomia*, 31(2): 123-126.

Wang Z, Ma Q L, Ma F Q, et al., 1995. Study on. the biological character of the host of *Cordyceps sinensis, Hepialus yushuensis*. *Gansu Agricultural Science and Technology*, (12): 38-40.

Wang Z, Ma Q L, Ma F Q, et al., 2001. Study on. the biological character of the host of *Cordyceps sinensis, Hepialus menyuanens*. *Gansu Agricultural Science and Technology*, (7): 38-39.

Yang D R, Li C D, Shen F R, et al., 1991. Studies on the reproductive behavior of *Hepialus baimaensis* Liang. *Zoological Research*, 12(4): 361-366.

Yang D R, Li C D, Shu C, et al., 1996. Studies on the Chinese species of the genus *Hepialus* and their geographical distribution. *Entomologica Sinica*, 39 (4): 413-422.

Yang D R, Shen F R, Yang Y X, et al., 1991. Biology studies and notes on a new species of the genus *Hepialus* from Yuannan, China. *Acta Entomologica Sinica*, 34(2): 218-224.

Yin D H, Chen S J, Li L, et al., 2004. Study on. the biological character of the host of *Cordyceps sinensis, Hepialus biru* in Tibet. *Special Wild Economic Animal and Plant Research*, 2:1-5.

Zhang G R, Gu D X, Liu X, 2007. A new species of *Hepialus* (Lepidoptera, Hepialidae) from China. *Acta Zootaxonomica Sinica*, 32 (2): 473 - 476.

Zhang S Y, Hu L Y, Wan Z G,1988. Studies on the bionimics of *Thitarodes armoricanus* oberthur. *Acta Entomologica Sinica*, 31(4): 395-399.

Zhao W Y, Yang D R, Shen F R, et al., 1989. Observation on the reproduction of the "insect herb" moth *Hepialus Yulongensis* Liang. *Acta Entomologica Sinica*, 32(3): 382-384.

Zhao Z H, 1992. Study on the main host swiftmoth, *Hepialus oblifurcus* Chu et Wang, of caterpillar fungus, *Cordyceps sinensis* (Berkeley) Sacc., in Kangding. *Acta Entomologica Sinica*, 35(3): 317-321.

Zou Z W, 2009. On the Insects of the Genus *Thitarodes* in Mt. Sejila of Tibet. Dissertation for Doctor Degree in Sun Yat-sen University, Guangzhou. pp.47-66.

Zou Z W, Liu X, Zhang G R, 2010. Revision of taxonomic system of the genus *Hepialus* (Lepidoptera, Hepialidae) currently adopted in China. *Journal of Hunan University of Science and Technology (Natural Science Edition)*, 25(1): 114-120.

In: Larvae: Morphology, Biology and Life Cycle   ISBN: 978-1-61942-662-7

Editors: Kia Pourali and Vafa Niroomand Raad   © 2012 Nova Science Publishers, Inc.

Chapter 9

SPAWNING AND NURSERY HABITATS OF NEOTROPICAL FISH SPECIES IN THE TRIBUTARIES OF A REGULATED RIVER

Maristela C. Makrakis[1], Patrícia S. da Silva[1], Sergio Makrakis[1], Ariane F. de Lima[1], Lucileine de Assumpção[1], Salete de Paula[1], Leandro E. Miranda[2] and João H. P. Dias[3]

[1]Grupo de Pesquisa em Tecnologia de Produção e Conservação de Recursos Pesqueiros e Hídricos - GETECH, Universidade Estadual do Oeste do Paraná, Toledo-PR, Brazil.

[2]US Geological Survey, Mississippi Cooperative Fish and Wildlife Research Unit, Mississippi State, MS, US

[3]Companhia Energética de São Paulo – CESP, Castilho-SP, Brazil

ABSTRACT

This chapter provides information on ontogenetic patterns of neotropical fish species distribution in tributaries (Verde, Pardo, Anhanduí, and Aguapeí rivers) of the Porto Primavera Reservoir, in the heavily dammed Paraná River, Brazil, identifying key spawning and nursery habitats. Samplings were conducted monthly in the main channel of rivers and in marginal lagoons from October through March during three consecutive spawning seasons in 2007-2010. Most species spawn in December especially in Verde River. Main river channels are spawning habitats and marginal lagoons are nursery areas for most fish, mainly for migratory species. The tributaries have high diversity of larvae species: a total of 56 taxa representing 21 families, dominated by Characidae. Sedentary species without parental care are more abundant (45.7%), and many long-distance migratory fish species are present (17.4%). Migrators included *Prochilodus lineatus*, *Rhaphiodon vulpinus*, *Hemisorubim platyrhynchos*, *Pimelodus maculatus*, *Pseudoplatystoma corruscans*, *Sorubim lima*, two threatened migratory species: *Salminus brasiliensis* and *Zungaro jahu*, and one endangered migratory species: *Brycon orbignyanus*. Most of these migratory species are vital to commercial and recreational fishing, and their stocks have decreased drastically in the last decades, attributed to habitat alteration, especially impoundments. The fish ladder at Porto Primavera Dam appears to be playing an important role in re-establishing longitudinal connectivity among critical habitats, allowing ascent to migratory fish species, and thus access to upstream reaches and tributaries. Establishment of Permanent Conservation Units in tributaries can help preserve habitats identified as essential spawning and nursery areas, and can be key to the maintenance and conservation of the fish species in the Paraná River basin.

INTRODUCTION

The Paraná River is the second longest river in South America, running through Brazil, Paraguay and Argentina over a course of some 3,089 kilometers (Agostinho and Júlio-Júnior, 1999), formed at the confluence of the Paranaiba and Grande rivers in southern Brazil. It is customarily divided into Upper, Middle, and Lower sections (Bonetto, 1989), each with distinctive geographic and biologic characteristics. The upper stretches are characterized by high human occupation and intense anthropogenic activities, and few areas are still in pristine conditions (Agostinho et al., 2007).

The diversity of fish is high in the Upper Paraná River basin, characterized by variety of reproductive strategies. Most species are sedentary or short-distance migratory, small and medium size with external fertilization, multiple spawning, short breeding periods and marked reproductive seasonality (Suzuki et al., 2004). Most large species migrate long distances to reproduce, and these species are considered the most important for commercial and recreational fisheries.

In the last decades the human interference in the Paraná River system has been considerable, and the most notable is the construction of dams (Agostinho and Gomes, 2002; Graça and Pavanelli, 2007), particularly on its upper reaches. Impoundments across the Upper Paraná River basin have had negative effects on natural fish populations for both sedentary and migratory species contributing to the diminished abundance and disappearance of fish species.

The fish community structure of a river depends on the integrity of the longitudinal connectivity of the system. The connectivity between river and floodplain or backwaters is essential in the life history of many fish (Lucas and Baras, 2001) that depend on seasonal use of flooded areas for spawning and feeding. Migratory fish in the Paraná River basin usually spawn in upper stretches of the large tributaries, and the lagoons are rearing areas in downstream sections of the tributaries of this river and along their borders and island (Agostinho et al., 2003; Nakatani et al., 2004). Thus, the blockage of the migratory routes with fragmentation of the environments and change from lotic into lentic conditions may induce local extinctions above barriers and reduce the fish populations downstream of those barriers (Makrakis et al., 2007a). The modification of the hydrologic cycle, attenuation of flooding and decrease of the reproduction areas and development of fish eggs and larvae are fundamental for that matter, though the intensity of the impacts depend on the location of the dam in relation to required habitats for different fish species.

Preservation or manipulation of critical spawning and nursery habitats can stabilize or increase the variable abundances and survival rates normally experienced by fish populations during early life, and thereby conserve or enhance fish populations (Fuiman and Werner, 2002). To better understand the population dynamic and the consequences of the environmental variability that affect the initial life stages of fish and recruitment in the Paraná

River Basin, we investigated if 1) fish species, primarily migratory fish, spawn in habitats of major tributaries to a large impoundment; and 2) tributaries have adequate spawning and nursery habitats to enable larval survival and development. To this end, spawning and nursery areas of fish species were studied over three spawning seasons (2007-2010) in the tributaries of the Porto Primavera Reservoir, in the heavily regulated Upper Paraná River, Brazil, identifying critical habitats for recruitment in this stretch of the basin where the information is still scarce. This chapter provides information that is relevant to the enhancement of early life stages of fish species expected to play an important role in the fisheries and and in the conservation of fish stocks in the Paraná River basin.

TRIBUTARIES OF PORTO PRIMAVERA RESERVOIR

A unique impounded reach in the Upper Paraná River is the Upper Paraná River Floodplain (Figure 1). It stretches from downstream of Porto Primavera Dam to the upper reaches of the Itaipu Reservoir, spans as wide as 20 km, especially on the western margin (Agostinho et al., 2003), and has large tributaries on the eastern margin. Flooded areas include active and semi-active channels, lagoons, elongated lowlands associated with paleochannels, and lowlands associated with the flood basin (Souza-Filho and Stevaux, 2004).

The Engenheiro Sergio Motta Hydroeletric Powerplant (known as Porto Primavera), belonging to Companhia Energética de São Paulo-CESP, is located in the main channel of Upper Paraná River, above the floodplain, along the border between São Paulo and Mato Grosso do Sul, Brazil (Figure 1). The dam is 11.4-km long and 22 m high, and has 16 surface spillways, with a discharge capacity of up to $53,600m^2.s^{-1}$ (Shibatta and Dias, 2006). A fish ladder (weir and orifice type) was built next to the dam, stretching 520 m to transcend the 19 m difference in elevations, allowing fish to reach the reservoir (Makrakis et al., 2007b). The Porto Primavera Reservoir drains a basin as large as $572,480$ km^2, and it has an area of 2.250 km^2.

Large tributaries are present on the east and west banks of the reservoir with various aquatic habitats: Aguapeí, Verde, Pardo and Anhanduí rivers. The tributaries have different characteristics, such as width, depth, substrate type, and preservation of riparian zones. The Pardo and Anhanduí rivers are similar, characterized by high mean discharge and water level, while the Aguapeí River stands out among the rest mainly by high specific conductance and turbidity. Verde River is characterized by high pH and dissolved oxygen.

The 305-km long Aguapeí River (Figure 1), located in São Paulo state, is on the eastern bank of the Paraná River (Paiva, 1982). It is characterized by a meandering pattern, with a width of 30 m. Its banks are composed of shrub and trees, as well as for aquatic macrophytes and several lagoons. A stretch of approximately 11.5 km was sampled, establishing five sites (RAG 1 and RAG 5), four in lotic and one in an oxbow lake (RG5) that provides continuous connection with the river. This lagoon has an area of approximately 2.0 hectares, with riparian vegetation preserved (Companhia Energetica de São Paulo - CESP, 2006).

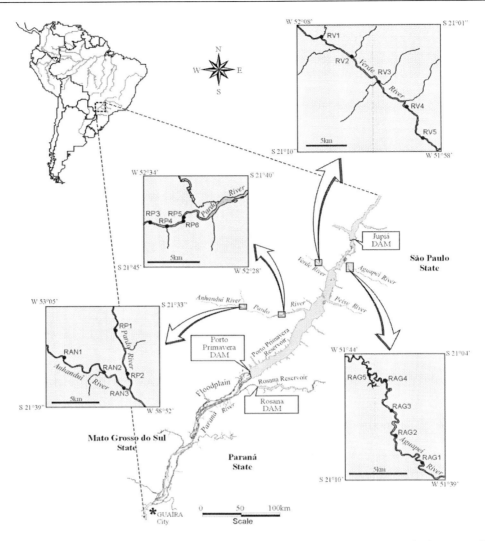

Figure 1. Sampling sites in the tributaries of Porto Primavera Reservoir, Upper Paraná River, Brazil.

The Verde River (Figure 1) is located on the western bank of the Parana River, covering an area of 3.3 km^2, part of the State of Mato Grosso do Sul (Paiva, 1982). The banks of this river are composed nearly pristine riparian vegetation. This river has many lagoons and wetlands areas. The study sites comprised five points (RV1 to RV5), all distributed in the main channel, totaling approximately 20km long.

The Pardo River (Figure 1) is located on the western bank of the Parana River, with an area of 35,050 km^2, in Mato Grosso do Sul State. The collections were conducted in two sections of the river totaling six different sampling sites. The first (upper section) presents only lotic sites (RP1 to RP2) located near the confluence area with the Anhanduí River covering a total of 9 km in length sampled. This section is totally free from the influence of the reservoir, has several areas of rapid waters and approximate width of 80 m. The riparian zone is well preserved, consisting of scrub and trees. Sites RP6 to RP3, in the lower section of the Pardo River, are located closer to the Porto Primavera Reservoir. The banks are composed of woody vegetation and shrubs and include several floating banks of aquatic macrophytes. The width between the margins is approximately 150 meters. The segment sampled comprised around 10 km, with three points from a lotic stretch, and one in a marginal lagoon (RP5) that provides permanent connection to the river. The main tributary of the Pardo is the Anhanduí River (Figure 1), which is 290-km long (Paiva, 1982). This site features a well-preserved riparian zone with shrubs and trees. The collections covered only lotic sites on this river with three sampling sites (RAN1, RAN2 and RAN3) distributed over 6 km of the river.

SPAWNING AND NURSERY HABITATS

Assessment of the temporal and spatial distribution of fish eggs and larvae in the tributaries of the Porto Primavera Reservoir, covering the Aguapeí, Verde, Pardo and Anhanduí rivers (Figure 1), were conducted monthly from October to March (reproductive period for major fish species from Paraná River Basin-Vazzoler, 1996) through 2007 to 2010, comprising three spawning seasons (1=2007/2008, 2=2008/2009, and 3=2009/2010). Ichthyoplankton samplings were conducted using a conical-cylindrical plankton net (0.5mm mesh) equipped with a flow meter for surface collections. Hauls were horizontal (20 cm deep) after nightfall (between 7 and 11 p.m.) during 10 minutes. Sampling included the main channel and marginal lagoons along the tributaries. Ichthyoplankton samplings were anesthetized with clove oil, fixed in buffered 4% formalin, and identified to the species level based on descriptions in Nakatani et al. (2001) and Graça and Pavanelli (2007). Species were also classified according to the reproductive strategies (Suzuki et al., 2004; Agostinho et al., 2003): MIG = long-distance migratory; SSC = sedentary without parental care; SCC = sedentary with parental care; SFIE = sedentary with internal fertilization and external development; and NC = species with unavailable information in literature.

Higher densities of fish eggs and larvae occurred between November and December, especially in 2009/2010 (spawning season 3; Figure 2). The fish reproduction period of major species from Upper Paraná River floodplain matches from October to March, and it is intensified by higher water temperature, longer days, and a rise in water level (Vazzoler, 1996); these factors trigger spawning. Water discharge is also a leading factor to trigger for reproduction of fish occupying the tributaries of Porto Primavera Reservoir; heavy rains contribute to the highest eggs abundance (in 2009/2010).

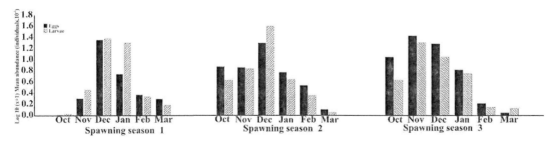

Figure 2. Mean abundance of fish eggs and larvae sampled in three spawning seasons (1=2007-2008); 2=2008-2009; 3=2009-2010), in the tributaries of Porto Primavera Reservoir, Upper Paraná River, Brazil.

The study tributaries were characterized by intense fish reproduction. High abundance of fish eggs and larvae occurs in all tributaries, especially in the Verde River (Figure 3). A longitudinal gradient has been reported in the distribution of eggs and larvae of fish species from the upper to the lower reaches of major tributaries of the Upper Paraná River; eggs have generally been more abundant at headwaters, decreasing toward the river mouth, and an opposite trend has been evidenced for larvae (Nakatani et al., 2004). Nevertheless, this pattern was not corroborated in the study tributaries as high abundances of fish eggs and larvae occurred throughout (Figure 4).

Spawning of most species occurs mainly in the main channel of tributaries, and lagoons are nursery habitats (Figure 3). Most fish that spawn in the main channel have pelagic eggs, efficient dispersion of eggs and larvae through flow, and larvae are placed in such a way as to facilitate their entry to nursery areas. However, species with demersal eggs prefer lagoons or shallow water to reproduce (Vazzoler, 1996), these eggs remain in the substrate adhered or not to marginal vegetation (Nakatani et al., 2001). Aquatic macrophytes in marginal lagoons provide favorable conditions for fish development due to abundant food and offer a wide availability of shelter. Cunico et al. (2002) state that the lagoons are true natural nurseries, supplying the needs of biological and ecological populations, such as breeding, feeding and growth.

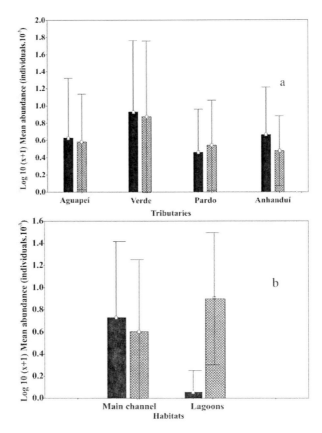

Figure 3. Mean abundance ± standard error of fish eggs (black bars) and larvae (gray bars) sampled in the tributaries (a) and the different habitats (b).

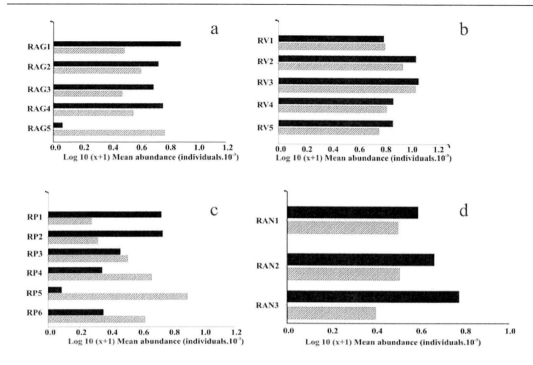

Figure 4. Mean abundance of fish eggs (black bars) and larvae (gray bras) in the different sampling sites of the tributaries: Aguapeí (a), Verde (b), Pardo (c), and Anhanduí (d).

LARVAE SPECIES COMPOSITION

The Upper Paraná River has high fish species diversity; about 310 fish species are known (Langeani et al., 2007), and they exhibit various reproductive strategies (Suzuki et al., 2004). In all, 56 species of fish larvae were found in the tributaries of Porto Primavera Reservoir, mainly Characiforms (45%) and Siluriforms (43%) (Chart 1). The tributaries show differences in species composition indicating greater reproductive activity of these species for a given tributary, and may be related to the characteristics of the site. The Verde River presents the greatest number of taxa, consequence of wetlands in this tributary, which benefit during the floods, making it natural nurseries for many fish species. Changes in the water level can provide optimal conditions for the larval development as new environments and shelter are flooded (Lowe-McConnell, 1987; Junk et al., 1989).

Most larvae species found in the tributaries are non-migratory species (short migrators or sedentary species-76%) (Figure 5). Many of these species have no parental care (45.7%-SSC), others have parental care (19.6%-SCC), some have internal fertilization and external development (8.7%-SFIE), and others have internal fertilization and internal development (2.2%-SFII). Sedentary species usually spend all their life cycle in lentic environments of the Upper Paraná River, such as oxbow lakes, lagoons and wetlands, and they do not depend of the rainy season to breed, because the conditions for the survival of offspring are suitable for most of the year.

Chart 1. Mean densities (larvae/10m^3) of larvae species in the tributaries of Porto Primavera Reservoir, Upper Paraná River, Brazil. SSC=sedentary species without parental care; SCC= sedentary species with parental care; SFIE= sedentary species with internal fertilization and external development; SFII= sedentary species with fertilization and internal development; NC=no available information; MIG = migratory species

Taxa	Tributaries				Strategies
	Aguapeí	Verde	Pardo	Anhanduí	
CHARACIFORM					
Parodontidae					
Apareiodon spp.	********				SSC
Prochilodontidae					
Prochilodus lineatus		********	********		MIG
Anostomidae					
Leporinus friderici	********				SSC
Leporinus spp.	********				SSC/MIG
Characidae	********		********		SSC
Astyanax altiparanae				********	SSC
Astyanax spp.	********	********			SSC
Bryconamericus spp.		********			SSC
Hemigrammus marginatus					SSC
Hemigrammus spp.					SSC
Moenkhausia aff. *sanctaefilomenae*		********	********		SSC
Salminus brasiliensis					MIG
Brycon orbignyanus	********		********	********	MIG
Serrasalmus spp.					SCC
Aphyocharax spp.	********		********		SSC
Roeboides descalvadensis		********	********		SSC
Serrapinus notomelas	********				SSC
Serrapinus spp.		********	********		SSC
Acestrorhynchidae					
Acestrorhynchus spp.	********		********	********	SSC
Cynodontidae			********		
Rhaphiodon vulpinus	********	********			MIG
Erythrinidae					
Hoplias spp.					SCC
Lesbisinidae					
Pyrrhulina australis			********		SSC
SILURIFORM					
Cetopsidae					
Cetopsis gobioides	********	********			NC
Callichthydae					
Hoplosternum littorale		********	********		SCC
Loricariidae					
Loricariichthys platymetopon	********	********	********		SCC
Pterygoplichthys anisitsi			********		SCC
Heptapteridae	********		********	********	
Pimelodella spp.			********	********	SSC
Rhamdia quelen	********	********	********		SSC

Taxa	Tributaries				Strategies
	Aguapeí	Verde	Pardo	Anhanduí	
Pimelodidae					
H. platyrhynchos/P. corruscans					MIG
Hypophthalmus edentatus				********	SSC
Iheringichthys labrosus			********		SSC
Pimelodus maculatus		*******			MIG
Pimelodus spp.			*******	********	SSC/MIG
Sorubim lima					MIG
Zungaro jahu				********	MIG
Doradidae					
Trachydoras paraguayensis			********		SSC
Auchenipteridae				********	
Ageneiosus inermis		********			SFIE
Auchenipterus osteomystax					SFIE
Parauchenipterus galeatus				********	SFIE
Tatia neivai					SFIE
GYMNOTIFORM					
Gymnotidae					
Gymnotus spp.					SCC
Sternopygidae					
Eigenmannia spp.		********			SSC
Apteronotidae					
Apteronotus spp.				********	SCC
SYNBRANCHIFORM					
Synbranchidae					
Synbranchus marmoratus		********	********		SCC
CYPRINODONTIFORM					
Poeciliidae					
Pamphorichthys sp. - "guaru"	********				SFII
PERCIFORM					
Cichlidae					
Geophagus spp. - "cará"			********		SCC

>1.0	0.50-1.0	0.11–0.49	0.02 – 0.10	******* >0<0.01

Of high importance is the occurrence of larvae of long-distance migratory species (17%-MIG) (Figure 6): curimba, *Prochilodus lineatus;* dourado, *Salminus brasiliensis*; piracanjuba, *Brycon orbignyanus;* dourado-cachorro, *Rhaphiodon vulpinus;* mandi, *Pimelodus maculatus*; bico-de-pato, *Sorubim lima;* pintado, *Pseudoplatystoma corruscans*; jaú, *Zungaro jahu*; and jurupoca/pintado, *Hemisorubim platyrhynchos/Pseudoplatystoma corruscans* (they are morphologically similar in the early stages of development). Among these species, *S. brasiliensis* and *Z. jahu* were already classified as threatened, *P. corruscans* as near threatened, and *B. orbignyanus* as endangered (high risk of extinction) (Abilhoa and Duboc, 2004). Long-distance migratory species generally spawn in the main channel, upstream from breeding habitats, and offspring drift downstream to reach nursery areas in lagoons, wetlands, and floodplain environments where they complete their development.

Larvae of migratory species occur in all tributaries of Porto Primavera Reservoir, but species composition varies, suggesting adults exhibit preference for tributaries (Figure 7). The highest abundance of migratory species are found in the Verde River, especially *B. orbignyanus*, *Z. jahu*, *S. brasiliensis*, and *H. plattyrhynchos/P. corruscans*. Comparing with other tributaries, this river seems to be the major holder of long-distance migratory species, because half of the species collected represent this behavior. Also, larvae of *P. lineatus* and *S. lima* occur mainly in the Aguapeí River. Higher abundances of *R. vulpinus* were found in the Pardo River and larvae of *P. maculatus* occurred especially in the Anhanduí River. Settlement patterns of habitat for some species reflect the suitability of an area for the survival of offspring in terms of food resources availability and refuge from predation (Arrington and Winemiller, 2006; Richardson et al., 2010).

Figure 5. Larvae of sedentary fish species of tributaries in Porto Primavera Reservoir, Upper Paraná River, Brazil. From top to bottom: *Hoplias* spp., *Apternotus* spp., *Gymnotus* spp., *Auchenipterus osteomystax*, *Loricariithys platymetopon*, and *Pterygoplichthys anisitsi*.

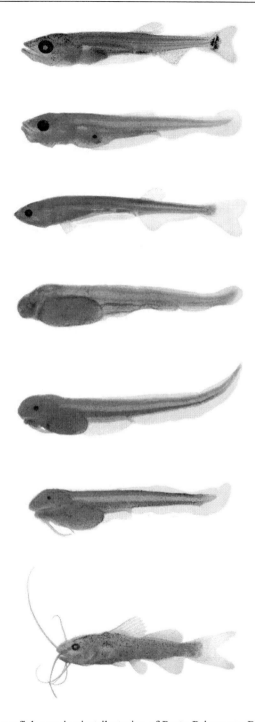

Figure 6. Larvae of migratory fish species in tributaries of Porto Primavera Reservoir, Upper Paraná River, Brazil. From top to bottom: *Brycon orbignyanus*, *Salminus brasiliensis*, *Rhaphiodon vulpinus*, *Prochilodus lineatus*, *Sorubim lima*, *Zungaro jahu*, and *Pimelodus maculatus*.

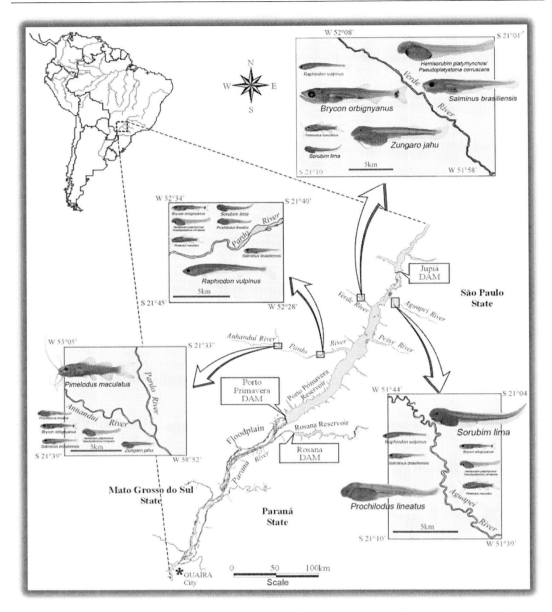

Figure 7. Distribution of the fish species larvae in the tributaries of Porto Primavera Reservoir, Upper Paraná River, Brazil. Larger larvae indicate higher abundances.

FINAL CONSIDERATIONS

The tributaries of Porto Primavera Reservoir in the Upper Paraná River have suitable conditions for reproduction and development of many fish species with diverse reproductive strategies, including the long-distance migrators. The reproductive activity of the species in the study tributaries, and especially the long-distance migrators, is comparable to that in the Upper Paraná River floodplain downstream Porto Primavera Dam, the last stretch of the

Upper Paraná River free of impoundments, and the least impacted riverine environment in the region.

This importance is confirmed by the occurrence of larvae of long-distance migrators of 9 of 19 possible species occurring in the Upper Paraná River. We suggest the fish ladder at Porto Primavera Dam may be playing an important role in re-establishing longitudinal connectivity among critical habitats, allowing ascent of migratory fish species into the reservoir, and eventually access to upstream reaches and tributaries. This result contradicts Pelicice and Agostinho (2008), who suggested fish passages into reservoirs act as ecological traps.

Establishment of Permanent Conservation Units in tributaries of Porto Primavera Reservoir may help preserve habitats identified as essential spawning and nursery areas, and may be key to the maintenance and conservation of the fish species in the Paraná River basin. Considering that the Paraná River is highly impounded, special attention should be given to the few remaining low-impact habitats as they continue to be targets of hydropower development that will likely intensify impacts on migratory fish stocks. An essential responsibility is to encourage reservoir managers and resource agencies to conserve these tributaries and avoid actions that potentially can condemn some fish species to extinction.

REFERENCES

Abilhoa, V., Duboc, L. F., 2004. Peixes. In Mikich, S. B. and Bernils, R. S. (Eds.). Livro vermelho da fauna ameaçada do Estado do Paraná. Pp. 581-677. Curitiba: Instituto Ambiental do Paraná.

Agostinho, A. A., Gomes, L. C. 2002. Biodiversity and fishery management in Paraná River basin: successes and failures. Universidade Estadual de Maringá, Maringá, Brazil: Blue Millenium, World Fisheries Trust, CRDI, UNEP. http:// www.unep.org/bpsp/Fisheries/Fisheries%20Case% 20Studies/ AGOSTINHO.pdf.

Agostinho, A. A., Júlio Jr., H. F. 1999. Peixes da bacia do alto rio Paraná. In: Lowe McConell, R. L. (Ed). Ecologia de Comunidades de Peixes Tropicais. Tradução de Vazzoler, A E. A M.; Agostinho, A. A.; Cunninghan, P. São Paulo. EDUSP. 456p.1999.

Agostinho, A. A., L. C. Gomes, H. I. Suzuki, Júlio Jr., H. F. 2003. Migratory fishes of the upper Paraná River basin Brazil. Pp. 19-89. In: Carolsfeld, J. B. Harvey, C. Ross and A. Baer (Eds.). Migratory fishes of South America: *Biology, Fisheries and Conservation Status.* Vitoria: World Bank, 372.

Agostinho, A. A., Pelicice, F. M., Petry, A. C., Gomes, L. C., Júlio Jr., H. F. 2007. Fish diversity in the upper Paraná River basin: habitats, fisheries, management and conservation. *Aquatic Ecosystem Health and Management*, London, 10 (2): p.174-186.

Arrington, D. A., Winemiller, K. O. 2006. Habitat affinity, the seasonal flood pulse, and community assembly in the littoral zone of a Neotropical floodplain river. *Journal of the North American Benthological Society,* 25 (1): 126–141.

Bonetto, A. A. 1989. The Parana' River system. In: Davies, B.R. and Walker, K.F., eds. The ecology of river systems. The Netherlands: Dr Junk Publ., pp. 541–556.

Companhia Energética de São Paulo - *CESP* - UHE Sergio Motta - Programa de monitoramento da ictiofauna e dos recursos pesqueiros. 2006. Levantamento de ovos e Larvas de peixes e reprodução - Relatório OA/031/2006.

Cunico, M. A., Graça, W. J., Veríssimo, S., Bini, L. M. 2002. Influência do nível hidrológico sobre a assembléia de peixes em sazonalmente isolada da planície de inundação do alto rio Paraná. *Acta Scientiarum*. 24: 383-389.

Fuiman, L. A., Werner, R. G. 2002. Fishery science: the unique contributions of early life stages. Oxford: Blackwell Science, 396 p.

Graça, W. J., Pavanelli, C. S. 2007. Peixes da planície de inundação do alto rio Paraná e áreas adjacentes. Maringá, EDUEM, 241p.

Langeani, F., R. M. C. Castro, O. T. Oyakawa, O. A. Shibatta, C. S. Pavanelli, And L. Casatti. 2007. Diversidade da ictiofauna do Alto Rio Paraná: composição atual e perspectivas futuras. *Biota Neotropica* 7(3): 1-17.

Lucas, M. C., Baras, E. 2001. Migration of freshwater fishes. Oxford: Blackwell Science Ltd.

Makrakis, M. C., Miranda, L. E., Makrakis, S., Xavier, A. M. M., Fontes-Júnior, H. M. and Morlis, W. G. 2007a. Migratory movements of pacu, Piaractus mesopotamicus, in the highly impounded river. *Journal of Applied Ichthyology* 1, 1- 18.

Makrakis, S., Makrakis, M. C., Wagner, R. L., Dias, J. H. P., Gomes, L. C. 2007b. Utilisation of the fish ladder at the Engenheiro Sergio Motta Dam, Brazil, by long distance migrating potamodromous species. *Neotropical Ichthyology*, 5: 197-204.

Nakatani, K., Agostinho, A. A., Bialetzki, A., Baumgartner, G., Sanches, P. V., Makrakis, M. C., Pavaneli, C. 2001. Manual de identificação de ovos e larvas de peixes brasileiros de água doce. Maringá, Eduem, 378p.

Nakatani, K., Bialetzki, A., Baumgartner, G., Sanches, P. V., Makrakis, M. C. 2004. Temporal and spatial dynamics of fish eggs and larvae. Pp 293-308. In: S. M. Thomaz, A. A. Agostinho and N. S. Hahn (eds.) The Upper Paraná River Floodplain: physical aspects, ecology and conservation. Backhuys Publishers, Leiden, The Netherlands.

Paiva, M. P. 1982. Grandes represas do Brasil. Brasília, Editerra, 302p.

Pelicice, F. M., Agostinho, A. A. 2008. Fish - Passage Facilities as Ecological Traps in Large Neotropical Rivers. *Conservation Biology*, 22: 180–188.

Richardson, D. E., Llopiz, J. K., Guigand, C. M, Cowen, R. K. 2010. Larval assemblages of large and medium-sized pelagic species in the Straits of Florida. *Oceanography,* 86: 8-20.

Shibatta, O. A., Dias, J. H. P. 2006. 40 peixes do Brasil: CESP 40 anos. Rio de Janeiro, Editora Doiis.

Souza-Filho, E. E. de, Stevaux, J. C. 2004. Geology and geomorphology of the Baía-Curutuba-Ivinheima River complex. In Thomaz, S. M., Agostinho, A. A. and Hahn, N. S., (Eds.). The Upper Paraná River Floodplain: physical aspects, ecology and conservation. Pp. 1-30. Leiden: Backhuys Publishers.

Suzuki, H. I., Vazzoler, A. E. A. M., Marques, E. E., Lizama, M. A. P., Inada, P., 2004. Reproductive ecology of the fish assemblages. In Thomaz, S. M., Agostinho, A. A. and Hahn, N. S., (Eds.). The Upper Paraná River Floodplain: physical aspects, ecology and conservation. Pp. 271-291. Leiden: Backhuys Publishers.

Vazzoler, A. E. A. M. 1996. Biologia da reprodução de peixes teleósteos: teoria e prática. Maringá, Eduem, 169p.

In: Larvae: Morphology, Biology and Life Cycle
Editors: Kia Pourali and Vafa Niroomand Raad

ISBN: 978-1-61942-662-7
© 2012 Nova Science Publishers, Inc.

Chapter 10

BIOFACTORIES. EXPRESSION OF RECOMBINANT PROTEINS IN INSECT LARVAE

Nicolás Urtasun, Alexandra M. Targovnik,
Laura E. Faletti, Osvaldo Cascone and María V. Miranda
Cátedra de Microbiología Industrial y Biotecnología. Facultad de Farmacia y Bioquímica.
Universidad de Buenos Aires. Junín,
Ciudad Autónoma de Buenos Aires, Argentina

ABSTRACT

In Biotechnology, the expression of recombinant proteins is a constantly growing field and different hosts are used for this purpose. Insects from the order Lepidoptera infected with recombinant baculovirus have appeared as a good choice to express high levels of proteins, especially those with post-translational modifications. Lepidopteran insects are extensively distributed in the world. Species like *Bombyx mori* (silkworm) have been explored in Asian countries to produce a great number of recombinant proteins for academic and industrial purposes. Several recombinant proteins produced in silkworms have already been commercialized. On the other hand, species like *Spodoptera frugiperda, Heliothis virescens, Rachiplusia nu* and *Trichoplusia ni* are widely distributed in the occidental world and Europe. The expression of recombinant proteins based on Lepidoptera has the advantage of its low cost in comparison with insect cell cultures. A wide variety of recombinant proteins, including enzymes, hormones and vaccines, have been efficiently expressed with intact biological activity. The expression of pharmaceutically relevant proteins, including cell/viral surface proteins and membrane proteins using insect larvae or cocoons, has become very attractive. This chapter describes the methods for scaling up the protein production using insect larvae as biofactories.

INTRODUCTION

In Biotechnology, there is a growing demand for fast and cost-effective production of biologically active eukaryotic biomolecules as proteins, glycoproteins, peptides and lectins.

Bacteria, yeast and mammalian cells are commonly used as hosts to produce recombinant proteins in short times, followed by insect cell cultures (Chapple et al., 2006; Kärkkäinen et al., 2009).

The baculovirus expression vector system is widely used, and, in most cases, cultures of the insect cell lines Sf9, Sf21 or High Five have been used as hosts.

The baculovirus-insect expression is accepted as an excellent system when an eukaryotic environment is required, for example in the production of proteins with biotechnological or pharmaceutical purposes. However, the main disadvantage of recombinant protein production in insect cell culture at industrial scale is its high cost (Ikonomou et al., 2003) because of the tissue-culture specialized facilities and reactors needed. Besides, at industrial scale, the risk of contamination is rather high.

A low-cost alternative is the production of recombinant proteins using live insect larvae as "biofactories". It is important to point out that the yield of protein is higher in larvae than in cultured insect cells. In this chapter, we describe the use of insect larvae as biofactories to scale up the expression of recombinant proteins.

The order Lepidoptera (butterflies and moths), the second largest order in the class Insecta, is a group of insects including more than 100,000 described species, many of which - considered destructive plagues during their larval stage - affect economically important crops such as soybean and corn. *Spodoptera frugiperda*, *Spodoptera littoralis*, *Trichoplusia ni*, *Helicoverpa zea*, *Heliothis virescens* and *Rachiplusia nu* are some of the most abundant and widely distributed lepidopteran species in the world.

Maeda and coworkers (1985) were pioneers in expressing the first recombinant therapeutic protein, human α-interferon, in silkworm (*Bombyx mori*) (Maeda et al, 1985). After that, the use of insect larvae as living factories for protein production has been explored as an alternative to cell-culture technology. Numerous enzymes (Medin et al., 1990; Tremblay et al., 1993; Loustau et al., 2008; Romero et al., 2010; Targovnik et al., 2010), antibodies (Reis et al., 1992; Gil et al., 2010), vaccines (Kuroda et al., 1989; Zhou et al., 1995; Perez-Filgueira et al., 2007; Perez-Martin et al., 2010; Millan et al., 2010), diagnostic proteins (Ahmad et al, 1993; Ismail et al., 1995; Katz et al., 1995; Barderas et al., 2000; Perez-Filgueira et al., 2006; Ferrer et al., 2007; Alonso-Padilla et al., 2010) and hormones (Mathavan et al., 1995; Sumathy et al., 1996) have been efficiently expressed in insect larvae using recombinant baculoviruses as vectors. Insect cells enable the overexpression of heterologous genes driven by strong viral promoters and provide essential components for the synthesis of properly folded and post-translationally modified complex proteins.

Baculoviruses are insect pathogens that regulate insect populations in nature and are being successfully used to control insect pests. Baculovirus-based insecticides are currently used worldwide. Many countries are developing new formulations and the most successful experiences in this field have been those with soybean in Brazil and with cotton in China. Currently, products containing wild type baculoviruses as active ingredients are accepted for sale. The industrial production of baculoviruses for insecticide purposes is generally performed directly in insect larvae.

Two species of the genus *Alphabaculovirus* are broadly applied in biotechnology as vectors to produce recombinant proteins in insect larvae: *Autographa californica* multiple nucleopolyhedrovirus (AcMNPV), which is by far the most widely used baculovirus expression vector, especially in American and European countries, and to a lesser extent

Bombyx mori nucleopolyhedrovirus (BmNPV), mainly adopted in China, India, Japan and other Asian countries.

AcMNPV infects a wide range of lepidopteran hosts (*S. frugiperda, T. ni, H. zea, H. virescens* and *R. nu*). Particularly, there is some industrial interest in the larvae of the cabbage looper moth, *T. ni,* which is an excellent host for AcMNPV and has been extensively used to produce a variety of recombinant proteins in biotechnology.

Nowadays, commercial scale protein production in *T. ni* larvae is available from several companies (Chesapeake PERL, Savage, MD and Entopath, Easton, PA.).

On the other hand, *B. mori* larvae have economic importance in silk production and for this reason the silkworm has been domesticated for thousands of years. The larvae have traditionally been cultivated on leaves of the mulberry *(Morus alba)*. Furthermore, in recent years, the silkworm has become an ideal multicellular eukaryotic model system for basic research. Although both the AcMNPV and BmNPV systems provide high level expression of foreign genes using larval hosts, the silkworm larva offers several additional advantages, i.e., it is easy to rear, it is large (120 mm in length in the last stage) and easy to manipulate, it has a relatively short life cycle (approx 7 weeks), and its genetics and biology have been well documented. In contrast, *S. frugiperda, H. zea, H. virescens* and *T ni* have five or six larval stages depending on the temperature and food, reaching only 35-40 mm in length. *R. nu* is the smallest, reaching a maximum length of 20 mm. *B. mori* larvae are not susceptible to AcMNPV, whereas *S. frugiperda, S. littoralis, T. ni, H. zea, H. virescens* and *R. nu* are not susceptible to BmNPV. All these species can be reared under laboratory conditions.

CONSTRUCTION OF THE RECOMBINANT BACULOVIRUS AND GENERATION OF VIRAL STOCK

Baculoviruses are rod-shaped DNA viruses that replicate in the nucleus of insect cells. Their double-stranded circular genomes are 80-180 kbp in size depending on the virus species. The complete genome sequences of 53 baculovirus isolates have been recently published in GenBank (van Oers, 2011).

Baculoviruses are characterized by having two different virion types: budded virions (BVs), which bud from the cell membrane and spread infection from cell to cell in an infected host insect, and a second type, called occlusion-derived virus (ODV), which is assembled entirely in the nucleus of infected cells and is occluded in large proteinaceous occlusion bodies (OBs).

Gene expression of baculoviruses occurs in four phases: immediate-early, delayed-early, late and very late (Passarelli and Guarino, 2007). In the very late phase, the virions are occluded, and two proteins, polyhedrin (29 kDa) and P10 (10 kDa), are produced in very large amounts. Polyhedrin forms the matrix of OBs in which the virions are embedded, while P10 is involved in the release of OBs from the nucleus of infected cells at the end of the infection.

Polyhedrin and P10 are needed to complete the infection cycle in a larval population via horizontal transmission but are not required to produce BVs, the form of the virus responsible for the systemic infection of the larva and for infection of insect cells in culture. As a consequence, the *polh* and *p10* promoters can be used to drive the expression of foreign genes in insect cells, the basis of the baculovirus insect cell expression system (Summers 2006).

The construction of a recombinant baculovirus is not achieved by direct cloning of the foreign gene because the baculovirus genome is very large to manipulate. So, the classical way of cloning is by homologous recombination between the viral genome and a transfer plasmid carrying the foreign gene under control of the *p10* and *polh* promoters. This recombination event may be carried out inside the insect cells by cotransfection with the transfer vector and viral genome or inside a bacterium and then transfecting insect cells with a bacmid purified from the bacterial culture.

The recombinant baculovirus construction, titration and cloning must be carried out in insect cell lines like Sf9, Sf21 or High Five in the case of AcMNPV or in BmN-4, BmN-5 cell lines in the case of BmNPV. It should be pointed out that most of the recombinant baculoviruses are constructed conventionally by replacing the polyhedrin gene (polh) with that of interest. This type of recombinant baculovirus is known as polyhedrin-minus genotype (polh-) because of its incapability to produce polyhedra and its low effectiveness to infect larvae by oral inoculation. The larva infection with polh- recombinant baculoviruses is restricted to the intrahemocelical injection of BV. On the other hand, oral infection using polyhedra is a simple methodology preferred when a large number of insects have to be inoculated, especially at large-scale production of recombinant proteins. Moreover, the oral-infection-based process is not a labor-intense or time-consuming technology, and does not require any degree of expertise in virus handling in comparison with the intrahemocelical infection strategy. Furthermore, different authors have demonstrated that there is no competition effect on protein synthesis when polyhedrin expression is driven by the *p10* promoter in larvae infected with polh+ recombinant baculovirus (Gatehouse et al., 2007; Romero et al., 2011).

A third strategy consists in obtaining baculovirus suspensions as preoccluded viruses (POV). POV are obtained when the polh- BVs begin to form occlusion bodies. POV are capable of infecting insect larvae by ingestion, but this strategy needs to develop complex methods based on the recovery of this type of virus from the nucleus of infected cells (Hughes and Wood, 1996).

INFECTION OF INSECT LARVAE WITH RECOMBINANT BACULOVIRUS

Lepidopteran pest insects such as *S. frugiperda* are permissive hosts to AcMNPV infection with BVs injected into the hemocele but resistant to oral infection (Haas-Stapleton et al., 2003; Haas-Stapleton et al., 2005). In contrast, *R. nu* oral and intrahemocelical susceptibility has been recently demonstrated (Ferrer et al., 2007; Romero et al., 2011).

Different host species have demonstrated some degree of developmental resistance, i.e. increasing resistance with age, although observed only when the virus is administered orally. Some species, such as *S. frugiperda* and *H. zea* are resistant to oral infection but they exhibit no systemic resistance to AcMNPV when budded virus is injected intrahemocelically. However, *H. virescens* is known to be a fully permissive host and little information exists about *R. nu* susceptibility (Ferrer et al., 2007; Romero et al., 2011). Resistance mechanisms result in various degrees of larval susceptibility to infection and this is decisive for choosing the infection route.

The optimization of protein expression in larvae is a complex task and one of the most important issues is the selection of the expression vector to be used. Another decision includes the larvae selection at the appropriate developmental stage. Larvae should be large enough to facilitate injection and yet not ready for pupae formation. Larvae should be selected between the 4th and the 5th instar stages, when protein synthesis capacity in the insect is optimal.

Larvae are injected subcutaneously into the hemocele with 5-50 µl recombinant baculovirus with a titer of about 1×10^7 pfu ml[-1] using a syringe with a 27.5-gauge needle. In some cases, it is recommended to place the larvae on ice for 15-30 min before injection. This procedure slows movement and makes the injection step easier. When larvae are orally infected, they are starved for 24 h and then fed a diet contaminated with polyhedra purified from insect cell cultures. It is important that all larvae consume the full diet. After infection, larvae are regularly fed a fresh non-contaminated diet until they are killed.

Infected larvae are maintained on an artificial diet and reared at 23-25°C in a 70% humidified chamber with a 16:8 photoperiod (L:D). A phenol-free diet is used for oral infection. Larvae of *H. zea* are very aggressive and need to be reared in isolation from each other but with *R. nu* or *T. ni* the larvae isolation can be avoided by offering abundant diet.

The time allowed for recombinant protein expression should be monitored in each case because it may vary with each particular protein. Generally a sample of hemolymph is enough to measure protein and make an expression curve. Usually 3-4 days are required to reach the peak of protein expression. It is important to inject a small number of insect larvae with recombinant baculovirus as a preliminary step in order to verify the protein expression in whole insects and select the optimal day for protein harvesting. Once this optimization step has been carried out and the protein quantified, the process can be linearly scaled up. The availability of automated rearing equipment and the fact that the larvae are nonallergenic to human handlers make scale-up and mass production of recombinant proteins very attractive for commercial protein production. With an automated facility for mass rearing and controlled conditions it is possible to scale up to kilograms of protein-containing larvae per week. The yield of the recombinant protein with the correct post-translational modifications can reach values ranging between micrograms and milligrams per larvae. In the case of recombinant horseradish peroxidase, the expression yields obtained in *R. nu* and *S. frugiperda* were 50 µg and 41 µg per larva respectively (Loustau et al., 2008; Targovnik et al., 2010), whereas that obtained in yeast was of 0.1 mg l[-1] and that in Sf9 cells 41.3 mg l[-1] (Morawski et al., 2000; Targovnik et al., 2010). In the last case, the cost of enzyme production was a hundred times more expensive than in a larval system (Targovnik et al., 2010).

The recombinant mouse anti-botulinum antibody fragment (Fab) has been expressed in *T. ni* larvae with a total yield of of 1.1 g/kg of larvae (O´Connell et al, 2007). Similar results were obtained with antigen protein-based virus-like particles (VLPs). Recently, Deo et al (2011) described the Rous sarcoma VLPs expression in silkworm larvae. The yield of the VLPs was approximately 8.2-fold higher than that obtained from stable cell lines. These results indicate that insect larvae are an interesting platform for large-scale application in vaccine development.

EXTRACTION AND PURIFICATION
OF RECOMBINANT PROTEINS

The downstream processing of recombinant protein of insect larvae has not been deeply studied. The variety of proteins which have been produced with this platform is so great that the development of standard methods is difficult. The choice of a particular downstream strategy is primarily dictated by the scale of operation, localization of the target protein and recovery yield. The aim is to separate contaminant proteins present in the host, viral proteins, DNA and viral particles. One of main difficulties is the high activity of proteases that may degrade the recombinant protein during sacrifice by homogenization. The time of harvest strongly influences the product quantity and quality. As the viral cycle progresses, the cells lyse and a significant amount of intracellular proteases and glycosidases appear in the homogenate. The first step is to achieve a clarified homogenate of insect larvae. This step is generally achieved by centrifugation or filtration to separate tissues and remove lipids. The clarified solution should have a yellow-green color. Besides, it is important to inhibit the melanization process during disruption to avoid protein yield loss.

Ion exchange and affinity chromatography have demonstrated to be efficient to purify different proteins from larval extracts (Nagaya et al., 2004; Romero et al., 2011).

CONCLUSION

Insect larvae infected with baculoviruses can serve as natural biofactories for the synthesis of proteins of interest in vivo. The high yield of protein reported (1-3 mg per larvae) in homogenates of 2 days postinfection at very low production costs make biofactories a very attractive alternative to traditional hosts like yeast or mammalian cells. However, further efforts must be made to improve the downstream processing of recombinant proteins.

REFERENCES

Ahmad S., Bassiri M., Banerjee A., Yilma T. (1993) Immunological characterization of the VSV nucleocapsid (N) protein expressed by recombinant baculovirus in Spodoptera exigua larva: use in differential diagnosis between vaccinated and infected animals. *Virology,* 192, 207-216.

Alonso-Padilla J., Jimenez de Oya N., Blazquez A., Loza-Rubio E., Escribano J., Saiz J., Escribano-Romero E (2010) Evaluation of an enzyme-linked immunosorbent assay for detection of West Nile virus infection based on a recombinant envelope protein produced in Trichoplusia ni larvae. *J. Virol Methods,* 166, 37-41.

Barderas M., Wigdorovitz A., Merelo F., Beitia F., Alonso C., Borca M, Escribano J. (2000) Serodiagnosis of African swine fever using the recombinant protein p30 expressed in insect larvae. *J. Virol. Methods,* 86, 129-136.

Chapple S., Crofts A., Shadbolt S., McCafferty J., Dyson M. (2006) Multiplexed expression and screening for recombinant protein production in mammalian cells *BMC Biotechnology,* 6, 49 DOI: 10.1186/1472-6750-6-49.

Deo V., Tsuji Y., Yasuda T., Kato T., Sakamoto N., Suzuki H., Park E. (2011) Expression of an RSV-gag virus-like particle in insect cell lines and silkworm larvae. *J. Virol. Methods.*, 177, 147-52.

Ferrer E., Gonzalez L., Martinez-Escribano J., Gonzalez-Barderas M., Cortez M., Davila I., Harrison L., Parkhouse R., Garate T. (2007) Evaluation of recombinant HP6-Tsag, an 18 kDa Taenia saginata oncospheral adhesion protein, for the diagnosis of cysticercosis. *Parasitol. Res.,* 101, 517-525.

Ferrer F., Zoth S., Calamante G. and Taboga O. (2007) Induction of virus-neutralizing antibodies by immunization with Rachiplusia nu per os infected with a recombinant baculovirus expressing the E2 glycoprotein of bovine viral diarrhea virus. *J. Virol. Methods,* 146, 424-427.

Gatehouse L, Markwick N, Poulton J., Young V., Ward V. and Chisteller J. (2007) Expression of two heterologous proteins depends on the mode of expression: comparison of in vivo and in vitro methods. *Bioprocess Biosyst. Eng.,* 31, 469-475.

Gil F., Perez-Filgueira M., Barderas M., Pastor-Vargas C., Alonso C., Vivanco F., Escribano J. (2010) Targeting antigens to an invariant epitope of the MHC Class II DR molecule potentiates the immune response to subunit vaccines. Virus Res., 155, 55-60.

Hass-Stapleton E., Washburn J., Volkman L (2003) Pathogenesis of Autographa californica M nucleopolyhedrovirus in fifth instar Spodoptera frugiperda. *J. General Virology,* 84, 2033-2040.

Hass-Stapleton E., Washburn J., Volkman L (2005) Spodoptera frugiperda resistance to oral infection by Autographa californica multiple nucleopolyhedrovirus linked to aberrant occlusion-derived virus binding in the midgut. *J. General Virology,* 86, 1349-1355.

Hughes P. and Wood A. (1996) In vivo production, stabilization and infectivity of baculovirus preoccluded virions. *App. Env. Microb.*, 62, 105-108.

Ikonomou L., Schneider Y., Agathos S. (2003) Insect cell culture for industrial production of recombinant proteins. *Appl. Microbiol. Biotechnol.,* 62, 1-20.

Ismail T., Yamanaka M., Saliki J., el-Kholy A., Mebus C., Yilma T. (1995) Cloning and expression of the nucleoprotein of peste des petits rumiants virus in baculovirus for use in serological diagnosis. *Virology,* 208, 776-778.

Kärkkäinen H., Lesch J., Määttä A., Toivanen P., Mähönen A., Roschier M., Airenne K., Laitinen O. and Ylä-Herttuala S. (2009) A 96-well format for a high throughput baculovirus generation, fast titering and recombinant protein production in insect and mammalian cells. *BMC Research Notes*, 2:63 doi: 10.1186/1756-0500-2-63

Katz J., Shafer A., Eernisse K. (1995) Construction and insect larval expression of recombinant vesicular stomatitis nucleocapsid protein and its use in competitive ELISA. *J. Virol. Methods*, 54, 145-157.

Kuroda K., Groner A., Frese K., Drenckhahn D., Hauser C., Rott E., Doerfler W., Klenk H. (1989) Synthesis of biologically active influenza virus hemagglutinin in insect larvae. *J. Virol.*, 63, 1677-1685.

Loustau M., Romero L., Levin G., Magri M., López M., Taboga O., Cascone O., Miranda M. (2008) Expression and purification of horseradish peroxidase in insect larvae. *Process Biochem.,* 43, 103-107.

Maeda S., Kawai T., Obinata M., Fujiwara H., Horuchi T., Saeki Y., Sato Y., Furusawa M. (1985) Production of human alpha-interferon in silkworm using a baculovirus vector. *Nature*, 315, 592-594.

Mathavan S., Gautvik V., Rokkones E., Olstad O., Kareem B., Maeda S., Gautvik K. (1995) High-level production of human parathyroid hormone in Bombyx mori larvae and BmN cells using recombinant baculovirus. *Gene,* 167, 33-39.

Medin J., Hunt L., Gathy K., Evans R., Coleman M. (1990) Efficient, low-cost protein factories: expression of human adenosine deaminase in baculovirus-infected insect larvae. *Proc. Natl. Acad. Sci. USA,* 87, 2760-2764.

Millan A., Gomez-Sebastian S., Nunez M., Veramendi J., Escribano J. (2010) Human papillomavirus-like particles vaccine efficiently produced in a non-fermentative system based on insect larvae. *Protein Expr. Purif.,* 74, 1-8.

Morawski B., Lin Z., Cirino P., Joo H., Arnold F. (2000) Functional expression of horseradish peroxidase in Saccharomyces cerevisiae and Pichia pastoris. *Protein Eng.,* 13, 377-84.

Nagaya H., Kanava T., Kaki H., Tobita Y., Takahashi M., Takahashi H., Yokomizo Y., Inumaru S. (2004) Establishment of a large-scale purification procedure for purified recombinant bovine interferon-tau produced by a silkworm-baculovirus gene expression system. *J. Vet. Med. Sci.,* 66, 1395-13401.

O´Connell K., Kovaleva E., Campbell J., Anderson P., Brown S., Davis D., Valdes J., Welch R., Bentley W., van Beek N. (2007) Production of a recombinant antibody fragment in whole insect larvae. *Molecular Biotechnol.,* 36, 44-51.

Passarelli A., Guarino L. (2007) Baculovirus late and very late gene regulation. *Curr. Drug Targets,* 8, 1103-1115.

Perez-Filgueira D., Resino-Talavan P., Cubillos C., Angulo I., Barderas M., Barcena J., Escribano J. (2007) Development of a low-cost, insect larvae-derived recombinant subunit vaccine against RHDV. *Virology,* 364, 422-430.

Perez-Filgueira, D., Gonzales-Camacho F., Gallardo C., Resino-Talavan P., Blanco E., Gomez-Casado E., Alonso C., Escribano J. (2006) Optimization and validation of recombinant serological tests for African Swine Fever diagnosis based on detection of the p30 protein produced in Trichoplusia ni larvae. *J. Clin. Microbiol.,* 44, 3114-3121.

Perez-Martin E., Gomez-Sebastian S., Argilaguet J., Sibila M., Fort M., Nofrarias M., Kurtz S., Escribano J., Segales J., Rodriguez F. (2010) Immunity conferred by an experimental vaccine based on the recombinant PCV2 Cap protein expressed in Trichoplusia ni-larvae. *Vaccine,* 28, 2340-2349.

Reis U., Blum B., von Specht B., Domdey H., Collins J. (1992) Antibody production in silkworm cells and silkworm larvae infected with a dual recombinant Bombyx mori nuclear polyhedrosis virus. *Biotechnology, (NY)* 10, 910-912.

Romero L., Targovnik A., Wolman F., Cascone O., Miranda M. (2011) Rachiplusia nu larva as a biofactory to achieve high level expression of horseradish peroxidase. *Biotechnol. Letters,* 33, 947-56.

Romero L., Targovnik A., Wolman F., Fogar M., Simonella M., Cascone O., Miranda, M. (2010) Recombinant peroxidase production in species of lepidoptera frequently found in Argentina. *New Biotechnology,* 27 , 857-861.

Sumathy S., Palhan V., Gopinathan K. (1996) Expression of human growth hormone in silkworm larvae through recombinant Bombyx mori nuclear polyhedrosis virus. *Protein Expr. Purif.,* 7, 262-268.

Summers M. (2006) Milestones leading to the genetic engineering of baculoviruses as expression vector systems and viral pesticides. *Adv. Virus Res.,* 68, 3-73.

Targovnik A., Romero L., Wolman F., Cascone O., Miranda M. (2010) Horseradish peroxidase production from *Spodoptera frugiperda* larvae: a simple and inexpensive method. *Process Biochem.,* 45, 835-840.

Tremblay N., Kennedy B., Street I., Kaupp W., Laliberte F., Weech P. (1993) Human group II phospholipase A2 expressed in Trichoplusia ni larvae-isolation and kinetic properties of the enzyme. *Prot. Expr. Purif.,* 4, 490-498.

van Oers M. (2011) Opportunities and challenges for the baculovirus expression system. *J. Invertebrate Pathology*, 107, S3-S15.

Zhou N., Zhang Y., Jing W., Li Z., Wu X. (1995) High expression of HBV 5 gene in Bombyx mori cell culture and in silkworms. *Chin. J. Biotechnol.,* 11, 149-156.

In: Larvae: Morphology, Biology and Life Cycle
Editors: Kia Pourali and Vafa Niroomand Raad

ISBN: 978-1-61942-662-7
© 2012 Nova Science Publishers, Inc.

Chapter 11

COMPARISON OF GROWTH, FATTY ACIDS AND AMINO ACID PROFILES DURING LARVAL ONTOGENY IN TWO MARINE FISH SPECIES

Margarida Saavedra[1] and Pedro Pousão-Ferreira

Instituto Nacional de Investigação Agrária e das Pescas (INIAP/IPIMAR-CRIPSul),
Olhão, Portugal

ABSTRACT

In aquaculture fish larvae rearing still represents a considerable challenge because fish larval nutrition is not yet fully understood. The initial larval stages still rely on live feed to survive and growth and live feed have substantial nutritional imbalances. It is crucial to determine the amino acids and fatty acid requirements of fish larvae in order to formulate suitable inert diets. It is possible to estimate larval amino acids requirements by determining the AA profiles from fish larvae carcass. Although this is just a first step it can give a preliminary indication of possible fish AA requirements. Also, if the AA profiles of fish are determined for several ages during fish larval ontogeny is possible to associate certain changes to specific events of fish development such as metamorphosis. Fatty acids are another extremely important nutrient during the first larval stages. Fatty acids are the most important energetic substrate and are crucial to the development of certain organs such as the eye. A change in the fatty acid profile of fish larvae may indicate a higher requirement of a specific fatty acid or its preferential spare. This study aims to compare the differences between the fatty acid and amino acid profiles of two different marine species and associate these changes to certain events occurring during fish larval ontogeny.

Keywords: Amino acids, fatty acids, growth, Diplodus sargus, Solea senegalensis

---

[1] E-mail address: margarida.saavedra@gmail.com, Tel.: +351 289 71 53 46, Fax.: +351 289 71 55 79.

1. INTRODUCTION

The knowledge on fish larval development has increased considerably over the last decades because of the expansion of aquaculture. Survival of fish larvae and its rely on live feed are still problems facing by the fish rearing industry.

Marine fish larvae undergo major morphological and cellular changes during the first month after hatched (Zambonino Infante and Cahu, 2001). It is expected, therefore, that these changes are reflected in differences on the amino acid (AA) and fatty acid (FA) profile of fish, especially if the morphological changes are pronounced.

Ten AA are considered indispensable for normal fish growth (Wilson, 1989). The composition of fish larvae is over 50% protein, in dry weight, and growth rate at this stage is very high, exceeding 50% body weight/ day (Conceição, 1997). Growth is mainly muscle protein deposition (Conceição et al, 1998) and AA are thought to be the major energy source during larval stages (Ronnestad et al., 1999) suggesting a high AA requirement during larval development stages. Growth rate is closely related to the protein supplied by the diet (Conceição et al., 2003). It is possible to maximize larval growth if the AA profile of the diet is as close as possible to the larval AA requirements (Akiyama et al., 1995, Conceição et al., 2003). Manipulations of AA composition can be very advantageous as they can increase both growth rates and food conversion efficiencies (Akiyama et al., 1995; Conceição et al., 2003). In rearing conditions, fish larvae AA requirements are supplied by rotifers and *Artemia* as they still rely on live feed to survive (Cahu and Zambonino Infante, 2001) due to the lack of a fully mature digestive tube, which incapacitates the full digestion of inert diets (Rønnestad *et al.*, 2003).

During larval stages occur several metabolic changes such as the development of several organs and digestive tract. These changes suggest AA requirements may change during fish larval ontogeny (Conceição *et al.*, 2003), especially when fish metamorphosis is pronounced (Aragão et al., 2004).

Another essential macronutrients, crucial for fish normal development, are the lipids. Lipids are a metabolic energy source and are involved in cell membranes and on the development of major organs (Sargent et al., 2002). Larvae of several marine fish species have shown to require n-3 highly unsaturated fatty acids (HUFA), particularly docosahexaenoic (DHA) and eicosapentaenoic acid (EPA). Fish larvae do not have the capacity to synthesize these fatty acids and in order to have a normal development and growth (Izquierdo *et al.* 1992) and they need to be provided by feed. HUFAs are cell membrane components (Sargent *et al.,* 1999) and major precursors of physiologically active molecules e.g., eicosanoids (Sargent *et al.* 1995; Sargent *et al.* 1999). HUFA's levels have been correlated to embryo and/or larva viability (Samaee *et al.* 2009). There is some evidence of a higher biological value for DHA instead of EPA during first life stages in species such as red seabream, gilthead seabream and turbot (Koven *et al.* 1993; Watanabe 1993; Rodríguez *et al.* 1994). This suggests that n-3 HUFA requirements may be related to the relative quantities of DHA and EPA and not just their individual quantities in the diet (Rodríguez *et al.* 1998).

White seabream, *Dipolodus sargus*, and Senegalese sole, *Solea senegalensis*, are two marine species that are being reared at small scale in countries such as Portugal and Spain. This study intends to evaluate the changes occurring in the AA and FA profiles during larval ontogeny of two marine fish species.

2. AMINO ACID PROFILES OF DIPLODUS SARGUS AND SOLEA SENEGALENSIS

The indispensable AA profile of fish carcass can be used as a first estimation of fish amino acids requirements (Wilson and Poe, 1985; Watanabe and Kiron, 1994). Using this AA profile it is possible to identify potential nutritional imbalances (Conceição et al., 2003).

The indispensable AA profile was estimated for two marine fish species that are being reared in captivity in Southern European countries such as Portugal and Spain. The AA profile was analysed at different stages of larval development.

Newly hatched *Diplodus sargus* were kept in 200 L at a density of 80 larvae/ L during 35 days. Larvae were collected for AA analysis at 2, 5, 12, 17, 25 and 35 days after hatched (DAH) (experimental details explain in Saavedra et al., 2006). From 2 to 35 days after hatched, the AA profile of *D. sargus* did not change significantly (Fig. 1). The AA showing highest relative content in the AA profile were lysine, leucine and arginine (Fig. 1). On the contrary, cystein, methionine and histidine showed the lowest relative AA percentage (Fig. 1).

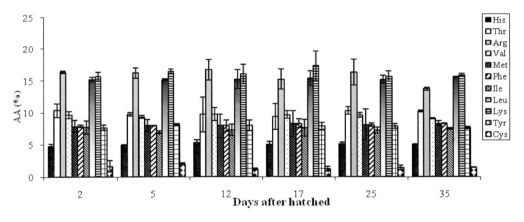

Figure 1. Indispensable amino acids profile (Mean and standard deviation) of *Diplodus sargus* from 2 days after hatched (DAH) until 35 DAH. IAA data are represented as a percentage of the IAA. His-histidine, Thr-threonine, Arg- arginine, Val- valine, Met- methionine, Phe- phenylalanine, Ile-isoleucine, Leu-leucine, Lys- lysine, Tyr- tyrosine, Cys- cystein. (Data from Saavedra et al., 2006).

Newly hatched Solea senegalensis were kept in 70 L fibreglass tanks at a density of 165 larvae / L. Larval samples were taken for AA analysis at 2, 5, 10, 16, 26 and 40 DAH (for details see Aragão et al., 2004).

The results showed the AA profile of Senegalese sole changed during larval ontogeny (Fig. 2). The differences in the AA profiles during ontogeny were reflected in changes on the relative content of most AA. Exceptions to this were methionine, histidine, isoleucine and cysteine, which were the only AA without significant changes during Solea senegalensis larval development (Fig. 2). Threonine and phenylalanine showed a higher percentage during earlier larval stages (2 to 10 DAH) compared to later post-larval stages (26 and 40 DAH). The opposite was observed for arginine and lysine. These AA had a lower relative content during the early larval development and its content increased in late ontogeny (2 and 10 DAH showed lower percentage compared to 16 and 40 DAH; 2, 5 and 10 DAH showed lower content compared to 16 and 40 DAH, respectively for arginine and lysine). Valine had lower percentage at 16 and 40 DAH whereas leucine had it at 5 DAH.

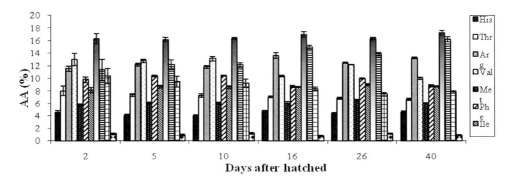

Figure 2. Indispensable amino acids profile (Mean and standard deviation) of *Solea senegalensis* from 2 days after hatch (DAH) until 40 DAH. IAA data are represented as a percentage of the IAA. His-histidine, Thr-threonine, Arg- arginine, Val- valine, Met- methionine, Phe- phenylalanine, Ile-isoleucine, Leu-leucine, Lys- lysine, Tyr- tyrosine, Cys- cysteine. (Data from Aragão et al., 2004).

3. FATTY ACID PROFILES OF DIPLODUS SARGUS AND SOLEA SENEGALENSIS

3.1. Diplodus sargus

First feeding larvae were reared in three 200 L conical cylindrical fibreglass tanks at a density of 100 larvae L⁻¹. Tanks worked in a semi-closed circuit and water temperature was 18 °C ± 0.6 °C, oxygen at 6.2 mg L⁻¹ ± 0.9 mg L⁻¹ and salinity at 37 ± 1 ppt. Water flow started at 0.6 L min⁻¹ and was later slowly increased to a maximum of 3 L min⁻¹ at 45 days after hatching (DAH). Photoperiod was 13 hours Light: 11 hours Dark. Feeding protocol consisted of rotifers (*Brachionus plicatilis*) enriched with Protein Selco (INVE Aquaculture, Belgium) from 3 to 20 DAH. At day 12, larvae were fed on *Artemia* nauplii (BE 480, INVE Aquaculture, Belgium), followed by *Artemia* metanauplii (*A. franciscana* Great Salt Lake, USA, INVE Aquaculture, Belgium) enriched with Super Selco (INVE Aquaculture, Belgium) from 17 DAH until 39 DAH. Dry feed (Nippai, Japan) was given from day 25 until the end of the experiment.

Fish larvae samples were taken at 0 (hatching), 2, 5, 8, 12, 17, 25, 35 and 45 DAH, consisting of 80 larvae initially and 20 larvae after 17 DAH. Larval samples were frozen in liquid nitrogen and then freeze-dried (RVT 400, Savant, NY). After weighted, all samples were analysed for fatty acid composition. Fatty acid composition was determined using the transesterification method by basic catalysis (Park *et al.* 2001). FAME were separated and quantified using a Varian CP 3800 gas chromatograph equipped with a flame ionization detector (250 °C) and a DB-WAX Polyetilene Glicol column (30 mx0.25 mm ID.x0.25 μm). Injector temperature was maintained constant at 250 °C over the 40 minutes of the analysis. The column was submitted to a temperature gradient, 5 minutes at 180 °C, an increase of 4 °Cmin⁻¹ for 10 minutes and, 220 °C for 25 minutes.

All statistical analyses were performed using the software Statistica® 6.0 (StatSoft Inc., Tulsa, USA). Homogeneity of variances was checked by Bartlett's test. When homogeneity of the variances was not achieved, data were subjected to the Kruskall-Wallis non parametric test.

The relative growth rate (RGR) started to be negative (between 0 and 2 DAH) but, after 2 DAH it showed a constant increase reaching the highest percentage at 17 DAH (20.78 % DW day $^{-1}$) (Table 1). After 17 DAH, RGR remained constant between 13 and 19 % DW day $^{-1}$.

Table 1. Relative growth rate (RGR) (% DW day $^{-1}$) of *Diplodus sargus* from 2 to 45 DAH

Age (DAH)	RGR (% DW day $^{-1}$)
0-2	-16.44
2-5	3.66
5-8	13.17
8-12	15.05
12-17	20.78
17-25	13.29
25-35	16.70
35-45	18.59

Table 2. Fatty acid composition (% Total FA) of white seabream *(D. sargus)* larvae. Different letters in the same row represent significant differences (p<0.05).[1] 22:6n3/20:5n3, [2] 20:5n3/20:4n6, R- Rotifers, NA-Artemia nauplii, AM- Artemia metanauplii, DF- Dry feed

Diet	R	R	R	R+AN	AM	AM+DF	DF
Larva Age	2 DAH (n=84)	5 DAH (n=84)	8 DAH (n=84)	12 DAH (n=84)	17 DAH (n=60)	25 DAH (n=40)	45 DAH (n=40)
14:00	1.75 ± 0.28	0.93 ± 0.24	0.56 ± 0.04	0.71 ± 0.03	1.6 ± 0.25	1.53 ± 0.52	1.45 ± 0.12
16:00	25.49 ± 1.45a	23.54 ± 1.41ab	18.23 ± 0.92b	16.77 ± 0.17b	20.46 ± 3.17ab	16.8 ± 0.64b	22.49 ± 0.87ab
18:00	6.88 ± 0.42a	9.43 ± 0.37c	11.16 ± 0.13d	10.78 ± 0.19d	9.00 ± 0.23bc	8.20 ± 0.33b	7.84 ± 0.20ab
Total - SFA	35.51 ± 2.06ab	35.56 ± 1.43ab	31.46 ± 1.12ab	29.76 ± 0.38b	32.99 ± 3.33ab	28.2 ± 1.07b	34.52 ± 1.24ab
16:01	4.21 ± 1.44ab	2.75 ± 0.31ab	2.27 ± 0.12b	2.69 ± 0.03ab	5.02 ± 0.22a	4.00 ± 0.64ab	2.82 ± 0.14ab
18:01	13.83 ± 0.35a	13.85 ± 0.89a	13.47 ± 0.25a	14.7 ± 0.12ac	23.48 ± 3.56b	21.44 ± 0.73b	17.70 ± 0.61c
Total - MUFA	19.36 ± 1.67a	18.01 ± 1.38a	17.48 ± 0.24a	19.93 ± 0.1a	31.3 ± 3.31b	29.27 ± 1.92b	22.33 ± 0.84a
18:2n6	4.72 ± 0.4ade	1.82 ± 0.09b	5.66 ± 0.17ce	7.37 ± 0.33c	6.25 ± 2.37ac	5.10 ± 0.18ade	3.20 ± 0.08bd
18:3n3	0.41 ± 0.01a	0.34 ± 0.01a	1.62 ± 0.04b	1.28 ± 0.09b	4.53 ± 0.12cd	5.51 ± 0.78c	1.70 ± 0.18b
20:4n6	1.64 ± 0.05ac	1.89 ± 0.12bcd	2.15 ± 0.02b	2.04 ± 0.01bcd	1.58 ± 0.21ad	2.08 ± 0.26bc	1.58 ± 0.03ad
20:5n3	5.31 ± 0.40a	5.05 ± 0.19a	5.55 ± 0.06a	6.31 ± 0.25a	5.27 ± 0.44a	10.82 ± 1.05b	12.42 ± 0.53b
22:5n3	1.06 ± 0.05a	0.97 ± 0.04a	1.57 ± 0.07ab	2.06 ± 0.13b	1.91 ± 1.03ab	1.98 ± 0.24ab	1.40 ± 0.06ab
22:6n3	25.59 ± 2.58ac	30.50 ± 2.43c	27.22 ± 1.25ac	23.58 ± 0.85acd	9.84 ± 3.09b	11.46 ± 0.08be	17.97 ± 1.38de
Total - PUFA	43.76 ± 2.96ab	44.09 ± 2.48ab	47.52 ± 1.49a	46.49 ± 0.13a	33.16 ± 6.85b	40.37 ± 1.87ab	41.39 ± 1.7ab
n-3 HUFA	31.96 ± 3.01bce	36.52 ± 2.65c	34.34 ± 1.37bc	31.94 ± 0.47bcd	17.01 ± 4.56a	24.69 ± 24.26ade	31.79 ± 1.97bce
DHA/EPA[1]	4.82 ± 0.17b	6.04 ± 0.28a	4.91 ± 0.18b	3.74 ± 0.28c	1.85 ± 0.43d	0.99 ± 1.07d	1.45 ± 0.05d
EPA/AA[2]	3.23 ± 0.15cd	2.67 ± 0.11de	2.58 ± 0.00e	3.09 ± 0.10ef	3.33 ± 0.17cf	5.03 ± 5.22a	7.84 ± 0.23b

Diplodus sargus larval fatty acids profiles showed significant differences between almost all larval ages (Table 2). Significant differences were found for 16:0. The relative content of this fatty acid decreased during larval ontogeny, having at 12 and 25 DAH lower levels compared to other ages (Table 2). Total monounsaturated fatty acids (MUFA) had a significant higher level at the ages of 17, 25 and 35 DAH. 18:3n3 showed lower relative contents at 2 and 5 DAH and higher relative content at 17, 25 and 35 DAH (Table 2).

Significant differences were also found for arachidonic acid (ARA). ARA had the highest percentage at 8 DAH (Table 2). At later stages of fish larvae development (25, 35 and 45 DAH), eicosapentaenoic acid (EPA) exhibited twice the content when compared to younger stages. Docosahexaenoic acid (DHA) had the highest content at 5 DAH and the lowest content at 17 DAH. DHA/EPA ratio was the highest at 5 DAH. This ratio decreased after 12 DAH and showed its lowest level at 35 DAH. Significant differences were also found for EPA/AA ratio that had higher levels at 35 and 45 DAH (Table 2).

3.2. Solea senegalensis

Newly hatched larvae were maintained in 300 L fiber glass tanks at a density of 15 larvae per liter. Senegalese sole was studied for 60 days. During this period, larvae were fed Artemia metanauplii enriched with Rich® during whole larval and postlarval phase. At 40 DAH, postlarvae started having dry feed (AgloNorse® 0, 1 and 2) in cofeeding with Artemia metanauplii. From 50 to 60 DAH, fish had dry feed alone. Fish larvae samples were collected at 19 DAH, when the pelagic larval stage ends, at 40 DAH, before weaning and at the end of the experimental period, 60 DAH (for details see Dâmaso-Rodrigues et al., 2010).

Table 3. Fatty acid composition (% Total FA) of Senegalese sole *(Solea senegalensis)*. AM- Artemia metanauplii, DF- Dry feed (Data from Dâmaso-Rodrigues et al., 2010)

Diet	AM	AM	DF
%	19 DAH	40 DAH	60 DAH
14:0	0.8± 0.2	0.6 ± 0.1	4.1 ± 0.4
16:0	13.3 ± 0.9	12.9 ± 0.7	15.6 ± 0.4
18:0	8.5 ± 0.9	8.6 ± 0.9	3.9 ± 0.4
Total - SFA	24.8 ± 1.7	24.1 ± 1.8	25.2 ± 0.5
16:1	6.7 ± 0.3	6.6 ± 0.2	2.4 ± 0.1
18:1	16.5 ± 0.7	17.1 ± 0.3	13.5 ± 0.2
Total - MUFA	29.3 ± 1.7	30.0 ± 0.5	34.6 ± 0.6
18:2n6	8.8 ± 0.6	8.9 ± 0.2	11.6 ± 0.4
18:3n3	10.9 ± 0.9	10.1 ± 0.4	2.0 ± 0.3
20:4n6	2.2 ± 0.3	2.2 ±0.2	1.0 ± 0.2
20:5n3	5.1 ± 0.7	4.2 ± 0.4	2.7 ± 0.1
22:6n3	7.3 ± 0.9	8.6 ± 1.1	14.3 ± 0.5
Total - PUFA	43.1 ± 1.9	43.3 ± 1.8	37.4 ± 0.4
DHA/EPA	1.4 ± 0.1	2.0 ± 0.3	5.3 ± 0.1

The data obtained in this study showed that the fatty acid profile of Senegalese sole did not change significantly from 19 to 40 DAH (Table 3). However, at 60 DAH, the relative content of 14:0 increased more than four times and 18:0 decreases to half. The total saturate FA remained approximately the same (Table 3). The percentage of MUFA increased slightly, especially due to an increase on the 20:1n-7. The total PUFA decreased at 60 DAH compared to 19 and 40 DAH because 22:5n-3 and 20:4n-6 were reduced to half but mainly because the relative content of 18:3n-3 was five times lower at the 60 DAH. Because the EPA decreased at this age, the DHA/EPA increased three times.

DISCUSSION

The results obtained from the fish carcass can only be considered as preliminary step to understand AA requirements (Wilson and Poe, 1985; Watanabe and Kiron, 1994) but they can provide useful information. The AA profiles of *Diplodus sargus* and *Solea senegalensis* are considerably different, indicating the AA requirements of these fish are possibly different as well. While *D. sargus* AA profile remained constant during larval development, the same was not observed for *S. senegalensis*. The differences between the AA profiles of the two species could be explained by their larval ontogeny. The changes observed in the AA profile during larval development of *S. senegalensis* are probably related to a pronounced metamorphosis which is marked by the transition from a pelagic stage to a final benthonic stage (Aragão et al., 2004). This transition involves an evident body change and eye migration (Aragão et al., 2004). On the contrary, *D. sargus* has a smooth metamorphosis and the transition from a post-larva to juvenile stage is not very pronounced and that might be the reason why its AA profile is considerably constant throughout larval development (Saavedra et al., 2006). Changes in the AA profile will possibly affect fish AA requirements.

Contrary to what was observed in the AA profiles, the fatty acid profiles of *D. sargus* changed considerably during larval development. It is possible to observe that often a change in the FA profile was coincident to a change in the diet. Similar observations were described for Atlantic halibut, *Hippoglossus hippoglossus,* where the fatty acids profiles from larvae reflected the fatty acid composition of the diet in most larval ages (Hamre *et al.* 2002). *D. sargus* larvae were fed for a considerable period on *Artemia* metanauplii enriched with HUFA but DHA relative content in the larvae was not as high as expected. In fact, the DHA content in the *Artemia* was a third of the content observed in rotifers and half of the dry feed. One of the main constraints to the use of *Artemia* is its low nutritive value but when the diet is supplemented with n-3 HUFA, growth rate is usually improved (Izquierdo 1996).

The highest growth rate observed in this trial was between 12 and 17 DAH. This is coincident to the period of *Artemia* nauplii. This is a curious result because during this period the percentage of DHA in the diet is the lowest. A possible explanation may be that although DHA relative content is low, the EPA percentage is higher compared to rotifers and ARA percentage is higher compare to both rotifers and dry feed. This suggests that fish larvae, between 12 and 17 DAH, may use preferentially other fatty acids rather than DHA for growth and development and/or the higher percentage of EPA and AA in *Artemia* nauplii is enough to increase growth rate in *D. sargus* larvae, considering larvae ingest more *Artemia* than rotifers at this stage.

At 25 and 45 DAH was observed an increase on larval growth rate. This suggests that fatty acid profile of dry feed may fulfil *D. sargus* fatty acid requirements at that age. At this stage the digestive tract is already fully developed which may be reflected in an enhancement on the ability to digest dry feed (Guerreiro *et al.* 2010).

The results obtained from *S. senegalensis* are consistent to the ones from *D. sargus*. A change in the fatty acid profile of *S. senegalensis* was also observed when the diet changed, from 40 to 60 DAH whereas when the diet was the same, at 19 and 40 DAH, the FA did not change significantly.

In conclusion, AA profiles during fish larvae ontogeny seem to be more related to the morphological and metabolic changes that the species go through during larval development rather than the diet provided. This was evident for both species studied which show different larval ontogenies. *S. senegelansis* with a pronounced metamorphosis showed significant changes in its AA profiles whereas *D. sargus* had a constant AA profile throughout its larval ontogeny as this species transition from the post-larva period to the juvenile stage is not pronounced. On the contrary, in both *D. sargus* and *S. senegalensis* the FA profiles were significantly affected by changes in the diet.

REFERENCES

Akiyama, T., Unuma, T., Yamamoto, T., Marcouli, P., Kishi, S. 1995. Combination of malt protein flour and soybean meal as alternative protein sources of fish meal in fingerling rainbow trout diets. *Fish. Sci.* 61, 828-832.

Aragão, C., Conceição, L.E.C., Fyhn, H.J. and Dinis, M.T. 2004. Estimated amino acid requirements during early ontogeny in fish with different life styles: gilthead seabream (*Sparus aurata*) and Senegale sole (*Solea senegalensis*). Aquaculture 233, 293-304.

Conceição, L.E.C., van der Meeren, T., Verreth, J.A.J., Evjen, M.S., Houlihan, D.F. and Fyhn, H.J. 1997. Amino acid metabolism and protein turnover in larval turbot (*Scophthalmus maximus*) fed natural zooplankton or *Artemia*. *Mar. Biol.* **129**, 255–265.

Conceição, L.E.C., Dersjant-Li, Y. and Verreth, J.A.J. 1998. Cost of growth in larval and juvenile African catfish (*Clarias gariepinus*) in relation to growth rate, food intake and oxygen consumption. *Aquaculture* 161, 95-106.

Conceição, L.E.C, Grasdalen, H and Ronnestad, I. 2003. Amino acid requirements of fish larvae and post-larvae: new tools and recent findings. *Aquaculture* 227, 221-232.

Dâmaso-Rodrigues, M.L., Pousão-Ferreira, P., Ribeiro, L., Coutinho, J., Bandarra, N.M., Gavaia, P.J., Narciso, L. and Morais, S. 2010. Lack of essential fatty acids in live feed during larval and post-larval rearing: effect on the performance of juvenile *Solea senegalensis*. *Aquacult. Int.* 18, 741-757.

Guerreiro, I., Vareilles, M., Pousão-Ferreira, P., Rodrigues, V., Dinis M.T. & Ribeiro, L. 2010. Effect of age-at-weaning on digestive capacity of white seabream (*Diplodus sargus*). *Aquaculture* 300 (1-4), 194-205.

Hamre, K., Opstad. I., Espe, M., Solbakken, J., Hemre, G-I. & Pittman, K. 2002. Nutrient composition and metamorphosis success of Atlantic halibut (*Hippoglossus hippoglossus,* L.) larvae fed natural zooplankton or *Artemia*. *Aquac. Nutr.* 8, 139-148.

Izquierdo, M.S., Arakawa, T., Takeuchi, T., Haroun, R & Watanabe, T. 1992. Effect of n-3 HUFA levels in Artemia on growth of larval Japanese flounder (*Paralichthys olivaceus*). *Aquaculture* 105, 73-82.

Izquierdo, M.S. (1996) Essential fatty acids requirements of cultured marine fish larvae. *Aquac. Nutr.* 2, 183-191.

Koven, W.M., Kolkovski, S., Tandler, A., Kissil G.Wm & Sklan, D. (1993) The effect of dietary lecithin and lipase, as a function of age, on n-9 fatty acid incorporation in the tissue lipids of *Sparus aurata* larvae. *Fish Physiology and Biochemistry* 10, 357–364.

Park, Y., Albright, K.S., Cai, Z.Y. & Pariza, M.W. (2001) Comparison of methylation procedures for conjugated linoleic acid and artefact formation by commercial (thimethylsilyl) diazomethane. *J. Agr. Food Chem.* 49, 1158 – 1164.

Rodríguez, C., Pérez, J.A., Lorenzo, A., Izquierdo, M.S. & Cejas, J. 1994. N-3 HUFA requirement of larval gilthead seabream *S. aurata* when using high levels of eicosapentaenoic acid. *Comp. Biochem. Physiol* 107, 693-698.

Rønnestad, I., Thorsen, A., Finn, R.N., 1999. Fish larval nutrition: recent advances in amino acid metabolism. Aquaculture 177, 201– 216.

Rønnestad, I., Conceição, L.E.C., Aragão, C. And Dinis, M.T. 2001. Assimilation and catabolism of dispensable and indispensable free amino acids in post-larval Senegal sole (*Solea senegalensis*). *Comp. Biochem Physiol. Part C, Toxicology and Pharmacology*, 130, 461-466.

Rønnestad, I., Tonheim, S.K., Fyhn, H.J., Rojas-García, C.R., Kamisaka,Y., Koven, W., Finn, R.N., Terjesen, B.F. and Conceição, L.E.C. 2003. The supply of amino acids during early feeding stages of marine fish larvae: A review of recent findings. *Aquaculture* 227, 147-164.

Saavedra, M., Conceição, L.E.C., Pousão-Ferreira, P. and Dinis, M.T. 2006. Amino acid profiles of *Diplodus sargus* (L., 1758) larvae: implications for feed formulation. *Aquaculture* 261, 587-593.

Samaee SM., Estévez A., Giménez G. & Lahnsteiner, F. 2009. Evaluation of quantitative importance of egg lipids and fatty acids during embryos and larvae development in marine pelagophil teleosts: with an emphasis on Dentex dentex. *Journal of Experimental Zoology* 311(10): 735-751

Sargent, J.R., Bell, J.G., Bell, M.V., Henderson, R.J. & Tocher, D.R. 1995. Requirement criteria for essential fatty acids. *Journal of Applied Ichthyology* **11,** 183–198.

Sargent, J., McEvoy, L., Estevez, A., Bell, G. Bell, M., Henderson J. & Tocher, D. 1999. Lipid nutrition of marine fish during early development: current status and future directions. *Aquaculture* 179, 217-229.

Sargent, J., Tocher, D., & Bell J. 2002. The lipids. In: J.E. Halver and D.M. Hardy, Editors, Fish Nutrition, Elsevier Science, USA pp. 181–257.

Watanabe, T. 1993. Importance of docosahexaenoic acid in marine larval fish. *Journal of World Aquaculture Society* 24, 152–161.

Watanabe, T. and Kiron, V. 1994. Prospects in larval fish dietetics (review). Aquaculture 124, 223-251.

Wilson, R.P. and Poe, W.E. 1985. Relationship of whole body and egg essential amino acid patterns to amino acid requirement patterns in chanel catfish *Ictalurus punctatus*. Comp. Biochem.Physiol., 80B, No2, 385-388.

Wilson, R. 1989. Amino acids and proteins. In: Halver, J. (Ed.), Fish Nutrition. Academic Press, San Diego, 11-151.

Zambonino Infante, J.L. and Cahu, C.L. 2001 Ontogeny of the gastrointestinal tract of marine fish larvae. Comparative Biochemistry and Physiology. Vol. 130, 477-487.

In: Larvae: Morphology, Biology and Life Cycle
Editors: Kia Pourali and Vafa Niroomand Raad
ISBN: 978-1-61942-662-7
© 2012 Nova Science Publishers, Inc.

Chapter 12

ADAPTIVE PLASTICITY AND EVOLUTION IN LARVAE WITH PARTICULAR REFERENCE TO THE CHIRONOMID MIDGE

Athol J. McLachlan[1]

School of Biology, Newcastle University, Newcastle upon
Tyne, NE17RU, UK

ABSTRACT

The theory of evolution is arguable the most influential theory shaping our thinking about biology. Yet it is almost exclusively a theory of adults. It virtually ignores the immature stages of organisms. Here I attempt to make a contribution toward redressing this imbalance, principally by reference to the aquatic larvae of a ubiquitous insect, the chironomid midge. These larvae occur in wide variety of inland waters, significantly including extreme habitats such as transient pools and hence provide good models to illustrate the adaptive response of larvae to a set of identifiable selective pressures. Importantly, the selective pressures experienced by larval stages are quite different to those experienced by adults and hence lead to divergent evolutionary trajectories.

INTRODUCTION

In this chapter I attempt to demonstrate the different adaptive universes inhabited by the larvae and adults of animals and discuss the degree to which the evolutionary trajectories of each are independent. This semi-independence holds despite the fact that larvae and adults share the same genome. I provide examples of larval adaptations in general and in particular among the salamanders and holometabolous insects and provide closer detail of the adaptive universe of the common chironomid midge. Instead of the traditional mutation centred approach to adaptation championed by others and notably by Dawkins (Dawkins 1999), I here

---

[1]Email:a.j.mclachlan@virgin.net.

adopt West-Eberhard's emphasis on phenotypic plasticity (West-Eberhard 2003). Phenotypic plasticity is defined by Wheeler (Wheeler 1910) as …"the power of an organism to adapt action to requirement without the guidance of a hereditary method of adjustment". This approach resolves many of the long-standing difficulties associated with the traditional approach. The concept of phenotypic plasticity is not new but has long been anticipated in popular culture in the legend of the werewolf (lycanthropy). Possibly the best know example of plasticity is the 'two legged coat effect' (West-Eberhard 2003), p53. This iconic example is the extraordinary morphological, anatomical and behavioral accommodations undergone by a goat born with only rear legs. Phenotypic plasticity includes the concept of modularity whereby the organism can be seen as modular in structure, each module of behavior, morphology or physiology, such as larva and adult stages in the life-cycle, being subject to *the rule of independent selection* (West-Eberhard 2003), p66. Switch points determine the timing of changes between alternative morphologies. The proximate mechanism of plasticity is the environmentally mediated response in gene expression. Note that it is the developmental stages, including larvae in the case of insects, which are flexible in this way and hence determine the outcome for the adult.

First an explanation of concepts. Epigenetics is the study of the flexibility of a genome's responses to environmental input, reviewed by (Petronis 2010). Depending upon environmental influences during development, genes can be either masked or expressed. Hence a single genome can produce dramatically different phenotypes as seen in the typically great phenotypic differences among the larvae and adults. Differences between embryo and adult human, tadpole and frog or caterpillar and butterfly are cases in point. Furthermore, facets of an anatomy, morphology, physiology and behavior can evolve semi-independently of each other. This phenomenon has been dubbed mosaic evolution by Myer, p51 (Myer 2002). There are other confounding effects, for example heterochrony. Heterochrony involves change in the timing of the appearance of features in the life-cycle. The best know example is the facultative neoteny of newts and salamanders; that is a shift in the timing of maturity resulting in the sexual maturity of juvenile forms.

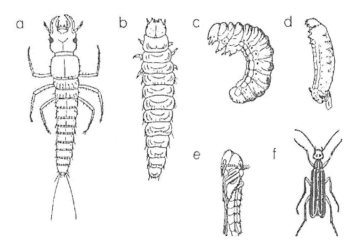

Figure 1. Divergent life-stage specializations in a beetle (*Epicauta vittata*). (a) First-instar triungulin larva, an active mobile egg hunter and predator. (b) Second-instar caraboid larva, a sedentary feeder; (c) Late-second instar scarabaeoid larva; (d) Immobile hibernating coarctate larva; (e) pupa; (f) adult. Not to scale. After (Riley 1878).

The case of salamanders is instructive because it involves an environmental parameter, an identifiable chemical, iodine in the water (Duellman and Trueb 1986), pp174, 179). The role of environmental iodine has lead to amphibia being central to studies of heterochrony and has resulted in spin-off for human medicine in the understanding of thyroid diseases such as cretinism (Swingle 1923; Venturi, Donati et al. 2000). In the same context, it has been suggested that many of the puzzling features of homo sapiens, such as hairlessness and the possession of a large cranium, can be understood as features of a neoteny (Gould 1977), pp356-358. Neoteny is among many possible consequences of changes in the timing of events fully discussed by Gould. The origin of the extraordinary larva of the holometabolous insect is a question that has long remained unresolved. The intriguing suggestion of Berlese (Berlese 1913), is that the insect larva is an embryo, freed from the confines of the egg to open a new world of ecological and adaptive opportunities. In other words the insect larva may provide another example of heterochrony. Insect metamorphosis has recently been developed into a comprehensive theory of development (Truman and Riddiford 1999).

The stark differences between immature stages of animals and their adults can hardly have escaped the attention of generations of biologists. Yet until recently, development was not brought into the main stream of evolutionary thinking. This is strange because the early work of such luminaries such as Haeckel and von Baer, upon which much of modern evolutionary biology is built, were concerned with just such studies; for review see (Gould 1977). Furthermore, both authors were well aware of the adaptive radiation of which the larva of a single individual is capable, enabling individuals species to exploit a variety of habits in the larva stage e.g.(Figure 1).

Since larvae and adult are so different, the transformation of one to the other is of special interest. In many species the change from zygote to adult, as in humans, is gradual but in others, as in the metamorphosis of holometabolous insects, it is abrupt. Metamorphosis involves a resting stage, the pupa, where the tissues of the larva are broken down and reorganized into those of the adult. This is an astonishing phenomenon, outstanding even among the many wonders of biology.

PHENOTYPIC PLASTICITY AND EVOLUTION

The link between phenotypic plasticity and evolution calls for an explanation. The core of the idea is as follows. Novel phenotypes that arise in response to environmental input do so because the external environment induces changes in gene expression. If there is a selective advantage to the modified genome it will be favored by natural selection and lead to a heritable novel genome. By the Buss hypotheses (Buss 1987), this happens during a brief widow of opportunity in the embryo before the sequestration of the germ line. Note that we are not here attempting to revive Lamarckism. This point is fully explored by Dawkins (Dawkins 1999). Importantly, the concept of phenotypic plasticity side-steps Weismann's barrier which prohibits the passing of information between phenotype and genotype. What I have attempted in this chapter is the briefest summary, fully explored by West-Eberhard in ground-breaking book already frequently alluded to in this chapter. Note that West-Eberhard removes mutation from its central position in evolution and puts the environment in its place.

CHIRONOMID MIDGES

Chironomid midges have a world-wide distribution and number many thousands of species. For this reason they early attracted the attention of a distinguished biologist as a suitable subject for the study of the principles of biogeography (Brundin 1965).The sole function of the adult midge is to mate and this is accomplished in a mating swarm (Downes 1969). Such swarms reflect a type of mating system less well known than that of birds and mammals and cries out for attention (Bradbury and Davies 1987). Chironomids are non-biting midges whose larvae, in terms of the proportion of time spent, dominate the life-cycle. Numerically too, the larvae dominate many fresh-water habitats and are therefore responsible for much of the energy flow in inland waters (Tokeshi 1995). For all these reasons, chironomids provide excellent opportunities for study. I now focus on some of the adaptations of the larvae to peculiarities of the fresh-water habitat.

PHENOTYPIC PLASTICITY AMONG CHIRONOMID LARVAE

The larvae of chironomid midges, particularly specialist invaders such as those in the genus *Chironomus*, show considerable ability to adjust life-cycle events to the peculiarities of the habitat such as ponds prone to unpredictable freezing (Bronmark and Hansson 2005), pp21,22, or evaporation (McLachlan and Ladle 2001). In these species two types of larvae exist, those that metamorphose early and those metamorphosing later. There is no evidence that these two types develop at different rates. Rather development rate appears to be fixed and some individuals spending longer in the larval stage resulting in larger adults. Differences in the time to metamorphosis shows as two discontinuous modes (Figure 2).

These two modes may have evolved as phenotype limited alternative options (ESS) of Maynard Smith, for review see (Krebs and Davies 1981), where those ecloding early benefit in a probability of metamorphosing before the habitat extinguishes but they carry a cost in the adult stage in terms of poor dispersal ability and limited eggs carrying capacity. The late mode by contrast, is risky for the larva but carries benefits for the adult in dispersal and fecundity. The size dimorphism is subtly cued to changes in the habitat and provides a nice example of adaptation.

The following events are most clearly seen in very small habitat patches such as puddles in the foot prints of large herbivores, fallen fruit, carrion or dung. In all of these, the habitat diminishes unpredictably due to evaporation or freezing or because it is consumed by the inhabitants. The result of a progressive reduction in the size of the habitat is a progressive crowding of inhabitants. It appears to be this crowding effect that triggers the appearance of late emergers. These have been dubbed giants by me earlier, in contrast to early emergers, the dwarfs (McLachlan 1989). There has been some doubt cast of ESS theory recently (West-Eberhard 2003), but see also (Dawkins 1999), p121 et seq. Nevertheless it is a beautiful idea with enormous explanatory power and I retain its useful attributes in this chapter. The giant/dwarf adaptation provides an interesting story that may illustrate principles widely applicable wherever the habitat has an unpredictable duration close to the minimum larva life-span.

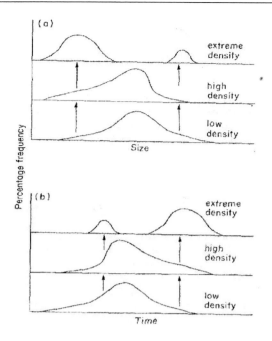

Figure 2. The original and subsequent size frequency distribution of phenotypes in a hypothetical population of puddle-dwelling *Chironomus* species subjected to increasing density. Arrows show the track of two phenotypes. (A) Body size. (B) Time to metamorphosis (from (McLachlan 1989).

A consequence of crowding is shortage of food and this leads to another story. There has been conjecture that the giants in the above example may gain by cannibalizing early emergers, the so called dwarfs, which like any animal provide a high protein food source (McLachlan 1989). Cannibalism has not been demonstrated among midges but is well know among rain-pool dwelling tadpoles and newts. For example, in the case of tadpoles of the frogs *Ceratophrys cornuta* and *Dendrobatis auratus*, giant tadpoles present under conditions of crowding undergo structural changes in jaw morphology appropriate to a carnivorous diet (Duellman and Trueb 1986) p.162. ESS theory requires that attention be given to what benefits accrue to the dwarfs in being consumed. One suggestion is that dwarfs gain in inclusive fitness by promoting survival of giant siblings (McLachlan 1989). For inclusive fitness theory to be applicable, any population of giants and dwarfs would have to be composed entirely of siblings. This is not an impossible condition but is rather restrictive and would be met only in the tiniest habitats such as those listed above. But these are common habitats, far outnumbering larger ones such as the ocean. Consequently, if they are viewed as islands, from the point of view of adaptive radiation they are of immense importance (McLachlan and Ladle 2001).

I turn now to an example of plasticity in the feeding biology of the larvae of chironomids. These larvae pass through four larval instars, the last three being mud dwellers, building tubes or silk and mud in which they pass their entire lives. Rhythmic undulations of the larval body force water through the tube to supply oxygen (Walshe 1951) and food, the latter as particles of organic matter in suspension (McLachlan and Ladle 2009). Plasticity in the structure of these tubes is interesting. What follows is a rethought and revised version of (McLachlan and Ladle 2009).The architecture of the tube of an iconic invader, *Chironomus plumosus* L., differ depending location in a lake or pond. Where sediment is shallow at the edges of the pond, tubes lie horizontally on the substratum and larvae in these tubes feed by grazing from the mud surface. In deeper sediments further into the pond, tubes are vertically orientated, and U-shaped with the two ends of the 'U' opening into the water. Larvae in such tubes filter- feed by spinning a net in the lumen of the tube. Light penetration is usually better at the edge of the pond and there are predictably more epipelic (mud surface) algae there, but in deeper water less light reaches the substratum and the principal source of food is suspended organic particles (McLachlan and Ladle 2009). The different tube-building strategies seem clearly related to the diversity of environments that chironomids inhabit (Figure 3).

Thus, tube shape plasticity in the larvae of chironomid midges such as *C. plumosus* seems adaptively appropriate to their location in the mud of a pond or stream. There are at least two possible explanations for the evolution of this phenomenon. An ESS with a fixed proportion of the two tube shapes in the population is one possibility. Here each might be genetically programmed to survive only if larvae settle in the appropriate depth of sediment, either shallow or deep. Such a lottery (Slatkin 1974), finds favor with Williams (Williams 1966), because it introduces genetical variety into the progeny of a organism so that the chance of some surviving in a novel environment are increased. However, a better solution, i.e. one carrying greater fitness benefits, would be for all larvae to inherit the ability to build their tubes and develop in whatever depth of sediment they find themselves. It is the second possibility that excites my interest here. In this case it is the environment, perceived as sediment depth, that appears to be the selective force leading to the evolution of behavioral flexibility enabling larvae to build different tubes as required. If this alternative proves to be correct, it as a pretty demonstration of the primacy of the environment over mutation in evolution. In the words of Lee van Valen , quoted by Stephen Gould p1."evolution is the control of development by ecology" (Gould 1977).

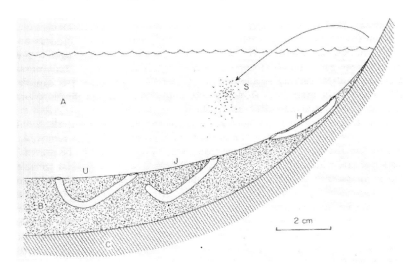

Figure 3. The distribution of the tubes of the larvae of *Chironomus spp.* A, water; B, sediment; C, hard underlay. S, sediment input due to erosion; H, horizontal tube; U, U-shaped tube; J, J-shaped tube (this type is not considered here). After (McLachlan 1977).

Plasticity is facilitated by modularity. To give just two examples, I consider first modularity of the larval form itself and secondly the case of larval imaginal disks. Some of the clearest examples of modularity among insects are see in differences in morphology and behavior among larval life-cycle stages. Rowe (Rowe 1985), gives the example of the damselfly *Xanthocneus zealandica* that exhibits a specialized territorial behavior in the early larval stages. Among chironomid midges similar larval stage modularity is known with the first larva instar being quite different in behavior and morphology to the remaining three instars (Figure 4).

First instar larvae are planktonic, evidently with the function of dispersal to suitable substrata on which to settle. The remaining three instars are essentially sedentary, building tubes of silk and mud at the bottom as described above. As can be seen in Figure 4, morphology too is often strikingly different depending on instar.

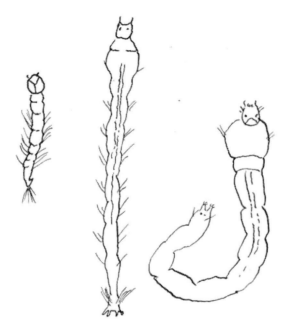

Figure 4. Larvae of a typical chironomid *Stenochironomus gibbus*, showing changes in morphology and ecology with development. From left to right; newly hatched pelagic first instar, later first instar boring in wood, later instar (not to scale). Redrawn after (Pinder 1995).

Figure 5. Leg and wing imaginal disks in the front end of a final instar *Chironomus* larva. Scale line, I mm. Redrawn after (Langton 1995).

Modularity of larval structure stems from the appearance, early in development, of clusters of specialized cells readily seen under a dissecting microscope. These are the imaginal disks (Figure 5), which eventually give rise to the adult structures such as wings, genitalia, etc (Nijhout 1991).

Development hinging on imaginal disks is quite different for the process of development seen in vertebrates, although that of vertebrates is also modular depending as it does on an hierarchical branching scheme (West-Eberhard 2003), p.57.

CONCLUSION

In this chapter I have attempted to set our some examples to illustrate the role on the larval stages of insects in the evolution of the adult phenotype. The theory of evolution is essentially a theory of adults (Buss 1987), yet it is in the immature stages that adult features and all evolutionary changes take place. Much depends upon developmental plasticity in the larva and the rule of independent selection. The adult is a static product of larval responses to environmental inputs. These environmental forces in turn mediate adaptive changes by the modification if gene expression. In the specific cased of the chironomid midge, a environment-centered view appears to help resolve some of the difficulties associated a traditional gene-centered view of evolution.

REFERENCES

Berlese, A. (1913). "Intorno alle metamorfosi degli insetti." *Redia* 9: 121-136.

Bradbury, J. W. and N. B. Davies (1987). Relative roles of intra- and intersexual selection. *Sexual Selection: Testing the Alternatives.* J. W. Bradbury and N. B. Davies. New York, Wiley.

Bronmark, C. and L.-A. Hansson (2005). *The Biology of Lakes and Ponds.* Oxford, Oxford University Press.

Brundin, L. (1965). "On the real nature of transatlantic relationships." *Evolution* 19: 496-505.

Buss, L. W. (1987). *The Evolution of Individuality.* Princeton, Princeto University Press.

Dawkins, R. (1999). *The Extended Phenotype.* Oxford, Oxford University Press.

Downes, J. A. (1969). "The swarming and mating flight of Diptera." *Annual Review of Entomology* 14: 171-297.

Duellman, W. E. and L. Trueb (1986). *Biology of Amphibians.* New York, McGraw-Hill.

Gould, S. J. (1977). *Ontogeny and Phylogeny.* London, The Belknap Press of Harvard University Press.

Krebs, J. R. and N. B. Davies (1981). *An Introduction to Behavioural Ecology.* London, Blackwell Scientific Publications.

Langton, P. H. (1995). The Pupa and Events Leading to Eclosion. *The Chironomidae.* P. D. Armitage, P. S. Cranston and L. C. V. Pinder. London, Chapman and Hall: 169-193.

McLachlan, A. J. (1977). "Some effects of tube shape on the feeding of Chironomus plumosus L. (Diptera: Chironomidae)." *Journal of Animal Ecology* 46: 139-146.

McLachlan, A. J. (1989). "Animal populations at extreme densities: size dimorphism by frequency dependent selection in ephemeral habitats." *Functional Ecology* 3: 633-643.

McLachlan, A. J. and R. Ladle (2001). "Life in the puddle: behavioural and life-cycle adaptations in the Diptera of tropical rain pools." *Biological Reviews* 76: 377-388.

McLachlan, A. J. and R. Ladle (2009). "The evolutionary ecology of detritus feeding in the larvae of freshwater Diptera." *Biological Reviews* 84: 133-141.

Myer, E. (2002). *What Evolution is.* London, Weidenfeld and Nicolson.

Nijhout, H. F. (1991). *The development and evolution of butterfly wing patterns.* Washington D.C., Smithsonian Institute Press.

Petronis, A. (2010). "Epigenetics as a unifying principle in the aetiology of complex traits and diseases." *Nature* 465: 721-727.

Pinder, L. C. V. (1995). Biology of the Eggs and First Instar Larvae. *The Chironomidae.* P. D. Armitage, P. S. Cranston and L. C. V. Pinder. London, Chapman and Hall: 87-106.

Riley, C. V. (1878). United States Entomology Commission. *First Annual Report.* . Washingotn , D.C., Government Printing Office. .

Rowe, R. J. (1985). "Intraspecific interactions of New Zealand damselfly larvae 1. Xanthocnemis zealandica, Ishnura aurora, and Austrolestes colensonis (Zygoptera: Coenagrionidae: Lestidae). ." *New Zealand Journal of Zoology.* 12: 1-15.

Slatkin, M. (1974). "Hedging ones evolutionary bets." *Nature* 250: 700-705.

Swingle, W. W. (1923). "Iodine and amphibian metamorphosis." *Biological Bulletin.* 45: 229-253.

Tokeshi, M. (1995). Production Ecology. *The Chironomidae. The biology and ecology of non-biting midges.* P. Armitage, P. S. Cranston and L. C. V. Pinder. London, Chapman and Hall: 269-292.

Truman, J. W. and L. M. Riddiford (1999). "The origins of insect metamorphosis." *Nature* 401: 447-452.

Venturi, S., F. M. Donati, et al. (2000). "Environmental iodine deficiency: a challenge to the evolution of terrestrial life? ." *Thyroid: official journal of the American Tryroid Association.* 10: 727-729.

Walshe, B. M. (1951). "The function of haemoglobin in relation to filter feeding in leaf mining chironomid larvae." *Journal of Experimental Biology* 28: 57-61.

West-Eberhard, M. J. (2003). *Developmental Plasitcity and Evolution.* Oxford, Oxford University Press.

Wheeler, W. M. (1910). *Ants.* New York, Columbia University Press.

Williams, G. C. (1966). *Adaptation and Natural Selection.* Princeton, Princeton University Press.

INDEX

A

access, xii, 154, 165
accommodations, 190
acetylcholinesterase, 41, 51
acid, xii, 6, 12, 13, 112, 115, 116, 118, 179, 180, 182, 183, 184, 185, 186, 187, 188
active site, 39
acute infection, ix, 84, 98
adaptation(s), 24, 28, 29, 31, 32, 108, 189, 192, 196
adaptive radiation, 191, 193
additives, 12
adenosine, 176
adhesion, 175
adipocyte, 120
adipose, x, 119, 121, 128, 129, 131, 132
adipose tissue, x, 119, 121
adjustment, 190
adults, vii, viii, xi, xii, 11, 37, 41, 45, 46, 48, 49, 88, 89, 96, 97, 98, 99, 100, 109, 120, 121, 133, 137, 139, 143, 146, 147, 148, 162, 189, 190, 191, 192, 196
advancements, 106, 110
adverse effects, 41
aetiology, 197
Africa, 41
age, ix, xi, 28, 30, 34, 57, 58, 61, 62, 63, 64, 65, 67, 70, 72, 74, 75, 79, 103, 137, 143, 172, 185, 186, 187
agencies, 165
aggregation, 141
agriculture, 40, 114, 135
algae, 11, 12, 194
alpha-fetoprotein, 39, 50
amino, xii, 39, 49, 110, 111, 112, 116, 118, 179, 180, 181, 182, 186, 187, 188
amino acid, xii, 39, 49, 111, 112, 118, 179, 180, 181, 182, 186, 187, 188

amphibia, 191, 197
amphibians, vii, 28
amylase, viii, 37, 48
anatomy, 109, 112, 113, 190
ancestors, 30, 120
ANOVA, 88
antibody, ix, 84, 85, 93, 98, 99, 100, 101, 102, 173, 176
antigen, 98, 100, 101, 173
antigenicity, 98
antitumor, 51
anus, 5, 129, 133
appetite, 10, 11
aquaculture, vii, x, xii, 1, 2, 13, 18, 19, 105, 106, 109, 110, 113, 114, 115, 117, 118, 179, 180
aquatic habitats, 155
Argentina, 154, 169, 176
arginine, 181, 182
arithmetic, 97
Artemia, 2, 7, 12, 13, 19, 109, 110, 111, 112, 115, 116, 117, 118, 180, 182, 183, 184, 186, 187
arthropods, vii, viii, 22, 23, 24, 25, 26, 27, 28, 31, 33, 34, 35, 36
ASEAN, 115
Asia, 41, 107, 118
Asian countries, xii, 169, 171
assessment, 111, 118
assimilation, 112, 118
athletes, 75
attachment, 38
Austria, 80
automation, 113
autonomy, 133
avoidance, 64, 79
axial skeleton, 4

F

J

K

L

M

Q

R